Silkworms and Science

by the same author

SCIENCE IN YOUR OWN BACK YARD

DISCOVERING CHEMISTRY

SCIENCE ON THE SHORES AND BANKS

MINERALS *(with Herbert S. Zim)*

Elizabeth K. Cooper

Silkworms and Science

The Story of Silk

Illustrated with photographs

Harcourt, Brace & World, Inc., New York

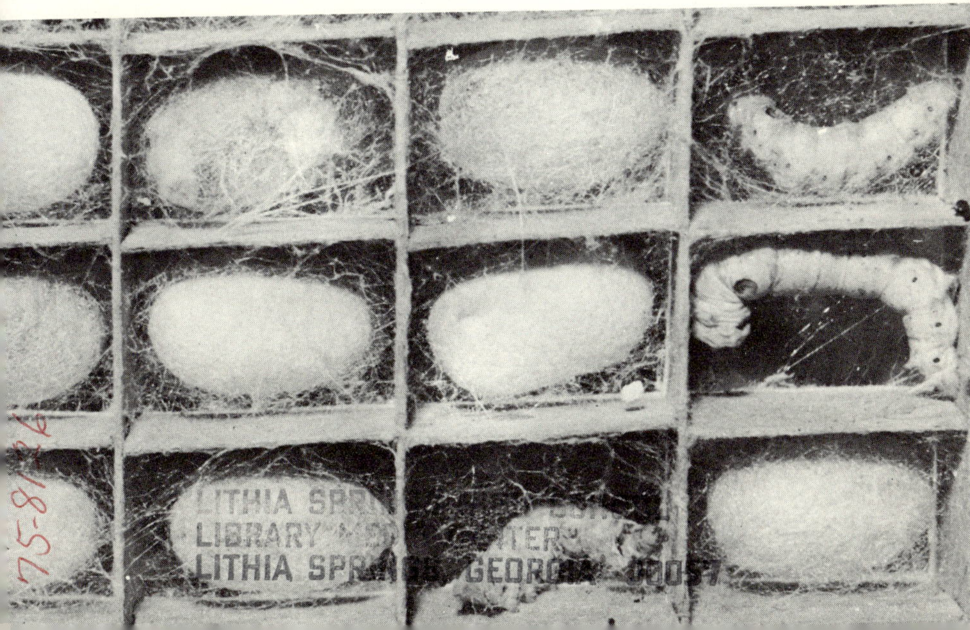

Library of Congress Catalog Card Number: 61-10109
Printed in the United States of America
'EFGHIJK

To children and teachers in the elementary schools of Santa Monica, California, where many generations of silkworms have spun cocoons, which we have unwound by hand

Acknowledgments

The author is indebted to the following organizations for their assistance in supplying information for the present volume:

The Japan Silk Association, Incorporated
Yurakucho, Chiyodaku
Tokyo, Japan

Japan Raw Silk Exporters Association
No. 1, Yamashitacho, Nakaku
Yokohama, Japan

The Silk Road Society
Shiba Takanawa
Tokyo, Japan

Japan Trade Center
393 Fifth Avenue
New York 16, New York

E.I. du Pont de Nemours and Company
Wilmington, Delaware

A special debt of gratitude is acknowledged to the two Japanese organizations that generously supplied most of the photographs used in this book, The Japan Silk Association, Inc., of Tokyo and Japan Raw Silk Exporters Association of Yokohama.

Appreciation is expressed to Bess Shirley King for her unique contribution as research assistant to the author, and to Ruth L. Kennedy for help in preparing the manuscript.

Elizabeth K. Cooper
Los Angeles, 1961

Contents

Silkworms and Science

Introducing the Silkworm and Its Favorite Food

The silkworm is an insect worth knowing. It is able to do something that no other insect can do. It can spin cocoons of silk that can be woven into silk cloth. It has been spinning silk for people to wear for nearly five thousand years. Today, in many countries all over the world, silkworms are spinning more silk than ever.

The little insect was first discovered in ancient China, where it fed on mulberry trees, stuffing itself with the shiny green leaves. Since then, silkworms have been carried from one continent to another and from country to country. Wherever they went, they had to have mulberry trees. In all the years, silkworms have never changed their eating habits.

The silkworm and the mulberry tree have a strange partnership. The mulberry tree can live very well without silkworms. But the silkworm cannot thrive or make good silk without the leaves of the mulberry tree. The history of the silkworm, then, is also the history of the mulberry tree. And what a long and exciting history it has been!

Silkworms feeding on mulberry leaves. *(Japan Silk Association, Inc.,* and *Japan Raw Silk Exporters Association)*

But perhaps even more exciting is the short natural life story of a single silkworm. This is something you will enjoy finding out about. You can raise some silkworms of your own. You can see them go through the different stages of their lives in a cardboard box in your own room. You can watch them as they spin their silken cocoons. And then, if you wish, you can learn how to unwind the cocoons and get some threads of pure silk spun by your own silkworms.

As you watch and work with your silkworms, you will be finding out some important things about the ways in which insects grow and change. You will also be learning how to observe living creatures in a scientific way.

Raising your own silkworms is a project you can do all by yourself, with a friend or two, or in school with all of your classmates working together. And once you begin, you can raise a new crop of silkworms every year.

How did it all begin? How did people ever find out how to get silk from silkworms? When and where was the discovery made? There are no proper scientific answers to these questions. In looking for possible answers, we have to go back to very ancient records left by the Chinese. And we have to go back to myths and legends that tell about prehistoric times — the days before any written records were made.

We find stories about the discovery of silk. Perhaps the stories have been made up by the poets of China. Perhaps the stories are true. Or it may be that the Chinese version of the discovery of silk is a mixture of true history and legend. You will have to decide for yourself what to believe.

1:
In the Emperor's Garden

The discovery of silk was made in faraway China, as long ago as 2640 B.C. There, since prehistoric times, silkworms had lived on the mulberry trees that grew wild on the hillsides. No one paid any more attention to them than to the other kinds of caterpillars that tattered the leaves with their feeding. Perhaps there were children who occasionally watched as a pearly white moth worked its way out of its silken cocoon.

But there were probably few if any adults who paid much attention to the caterpillars on the mulberry trees or to their fat shiny cocoons. And then, one day, someone became curious. It was Si-Ling-Chi, the young and beautiful wife of Huang-Ti, the emperor.

It was early spring, but the day was warm. Si-Ling-Chi and her ladies in waiting were sitting in the garden inside the palace walls. The sun was hot, and she and her ladies moved under the wide-spreading branches of a mulberry tree. As she sat there fanning herself on that hot afternoon, Si-Ling-Chi began to watch some silk-moth caterpillars spinning cocoons on one of the low branches. She saw the way in which the tiny animals spun out yards and yards of lustrous silk and wrapped it about their bodies. Fascinated, she watched the spinners until the sun dipped low in the sky and the chill of evening drove her back into the palace.

All night she kept thinking of the silkworms and their delicate strands of shiny silk. The next day, and for several days thereafter, she and her ladies returned to their place under the mulberry tree. They gazed at the fat, lustrous cocoons anchored to the branches of the tree. They saw them shimmer in the sunlight. How beautiful, thought Si-Ling-Chi, would be a royal gown woven from such airy strands! Such a gown would indeed be fit for an emperor – or even for a god. A plan began to form in her mind.

On the fourth day, when Si-Ling-Chi went into the garden, there were no more caterpillars on the mulberry tree. In their

The fat, creamy white cocoons. *(Japan Silk Association, Inc.,* and *Japan Raw Silk Exporters Association)*

places hung creamy white and golden cocoons. With the help of her ladies, she plucked the cocoons from the branches and carried them into the palace. There she examined the cocoons carefully. She turned them over and over in her hands. Perhaps she poured hot water into one of her fine porcelain basins and washed the cocoons. In any case, she finally found the end of one of the silken strands and began to unwind a cocoon. The ladies in waiting were soon unwinding cocoons, too. They worked for days, winding silk on reels made from twigs cut from the mulberry tree.

They found that the strand from a single cocoon was fairly strong, but far too fine to be woven on a loom. And so they began to unwind three, four, or five cocoons at one time. The separate strands were kept together by passing them through Si-Ling-Chi's golden ring. One lady held the ring as another twirled and twirled the twig on which the silk thread was being reeled.

When all of the silk had been taken from the first group of cocoons, Si-Ling-Chi and her ladies wandered all over the royal gardens gathering more and more cocoons. They noticed that the cocoons that yielded the silken strands were to be found only on the branches of mulberry trees. And so they went from one mulberry tree to the next, gathering the cocoons and bringing them back to the palace to be unwound.

Perhaps in their work, they discovered that the cocoons could be most easily unwound if they were first put into boiling water. Perhaps they learned to keep the cocoons bobbing in a metal or earthen pan on top of a small charcoal fire while they were being unwound. Perhaps they discovered that the tiny animal inside the cocoon had to be killed to keep it from breaking out of the cocoon and spoiling the silk. Perhaps. No one knows for sure.

But, according to the Chinese legend, Si-Ling-Chi did learn the secret of unwinding the cocoons of the silk moth. And then she learned to weave beautiful fabric from the silk. The first piece was made into a robe for Huang-Ti, the emperor. With her own hands, she embroidered it with delicate flowers and scrolls and green leaves. She sewed in the designs with strands of silk, which she had dyed all the colors of the rainbow. When the robe was finished, Si-Ling-Chi presented it to the emperor.

Huang-Ti was overcome with the beauty of the garment. Never had anyone, not even an emperor, worn anything so fine. He was both proud and excited as he asked Si-Ling-Chi to tell him where and how she had found the lustrous fabric that was as smooth as the petal of a rose and as light as a spider web. With many bows and blushes, the young wife of the emperor told about the strange little animals that hung their bundles of silk in the mulberry trees that grew in the palace gardens.

As the emperor passed his long, sensitive fingers over the silken folds of his new robe, he made a decision. Here in the royal palace they would capture some of the little spinners of silk. They would learn how to raise them. They would get many, many cocoons. And all the ladies of the palace would learn how to unwind the silken strands and weave them into garments for the royal family, and into banners and robes to honor the gods.

And these things they did. Each spring Si-Ling-Chi and her ladies tended the silkworms, collected the cocoons, reeled the silk, and finally wove the precious strands into shimmering fabric. After her death, Si-Ling-Chi became known as the Goddess of Silk. Temples were built in her honor.

Generations later, the rearing of silkworms and the reeling of silk were still the responsibility of the royal ladies who lived in the emperor's palace. When the first flicker of green began to show on the bare branches of the mulberry trees, the ladies in waiting were relieved of their regular duties. Some were sent out to gather the swelling buds and young leaves. The buds were split open. The tender, unopened leaves were fed to the newly hatched silkworms. Tiny leaves that had just burst forth were fed to silkworms that were a day or two old. As the silkworms grew in size and appetite, larger and larger leaves were gathered to feed them.

After the cocoons had been spun, all of the ladies of the court worked together reeling the silk and weaving it into fabric. On a certain day, the empress herself, dressed in her finest silk gown, went to the temple bearing offerings to honor Si-Ling-Chi, empress, goddess, mother of the silkworms.

How much of this really happened, we cannot say. We do know that the story of Si-Ling-Chi has been handed down from one generation to the next. And we do know that woven silk was worn in China by those of royal blood thousands of years ago. It may well be that silk had its beginnings in the ancient palace gardens of the emperor Huang-Ti, over 4,600 years ago. The story of exactly how it happened is lost in legend.

The oldest Chinese books that we know anything about mention the raising of silkworms and mulberry trees by the emperor and his nobles. In the springtime, according to these ancient accounts, all the ladies in the palace drew lots. The winners were permitted to work in the silkworm nurseries. It seems that, from the beginning, the production of silk was a special activity of the emperor's court.

Silk cloth, of course, was both rare and costly. Only those of royal or noble birth were permitted to wear it. In those ancient times, silk was found only among the ruling families and in the temples.

Very early, silk had special religious uses. The Chinese temples that honored the spirits of earth and sky were hung with shining silk banners. The Temple of Heaven was hung with blue silk. The Temple of Earth was done in yellow. The Temple of the Sun was in red, and the Temple of the Moon, in white. What a bright and shining sight it must have been to see the emperor and empress and their lords and ladies, each in brilliant silken robes, making a procession into a silk-hung temple!

Silk became a symbol of royalty, of culture, of wealth and power. Artists made paintings on delicate panels of silk. Poets wrote their verses on silk scrolls. Once the Chinese knew how to get the silk strands from the silkworms, they had plenty of silk for use in their own country and also to trade. The silk cloth that was traded outside of China brought in much gold. The emperors of China did not want other countries to find out how to grow their own silk.

The production of silk, which is called sericulture, was the shining secret of ancient China. The secret was well kept for *several thousand* years!

2:
The Ancient Silk Road

There was a law in ancient China that helped to keep the secret of silk inside the country. According to this law, anyone carrying the eggs of the silk moth or the seeds of the mulberry tree out of China would be killed. Everything about the rearing of silkworms and the unwinding of cocoons was the special concern of the emperor. Silk was one of the most important products in the ancient world, and the Chinese emperor had no intention of letting people in other countries know how silk was made.

Travelers and traders came to ancient China, crossing mountains and deserts and dangerous seas in order to see its riches and buy from its treasures. There were jewels and furs and spices. There were rare woods and carvings of unbelievable delicacy. There were exotic perfumes. Most of all, there was silk, which was made nowhere in the world except in China.

So great was the desire of men everywhere for the rich, lustrous fabric that daring traders from many lands made the long journey to China. By the year 126 B.C., there was a regular trading route, which has since been called the "Silk Road." It was a route but in no way a road. It led through rocky passes and across parching deserts, where camels were the only means of transportation. It led through lands held by wild tribes, where the fierce Mongols swept down upon the silk caravans and robbed and murdered. It led across bleak, empty country, where lack of food and water was a constant danger. But, by the time the traders reached their own lands – if they were lucky enough to finish their journey at all – the silk cloth they carried was worth literally its weight in gold.

And so, year after year and century after century, traders traveled the Silk Road. This six-thousand-mile route was the only tie between China in the east and Greece and Rome to the west. Silk was so rare and so expensive outside of China that even kings and emperors could afford very little of it.

The ancient Silk Road. *(Japan Silk Association, Inc.,* and *Japan Raw Silk Exporters Association)*

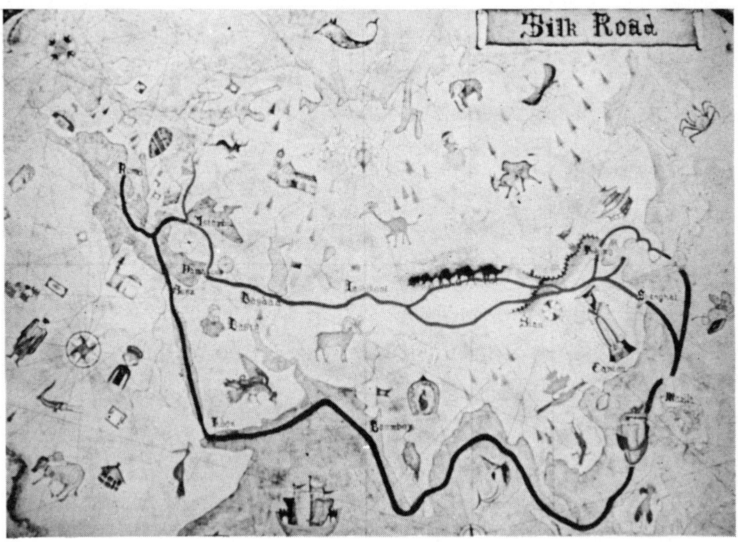

And when a piece of silk wore out, it was slowly taken apart, thread by thread. The broken threads were tied together and woven into something new.

Along with the silk, traders brought home stories of the fabulous riches of China, the home of the precious and mysterious fabric. Rulers of other countries knew that silk had much to do with China's great wealth. They envied China its secret knowledge. They wanted to discover the secret of silk. But, try as they might, they could not find out. They had no idea what made silk.

Today we may wonder how a whole country could keep a secret for thousands of years. Though the Silk Road was traveled by traders and others, there could not have been many visitors in China at any one time. And whenever a caravan or group of travelers arrived, the emperor and his court were probably notified at once. It would not have been difficult, then, to have all strangers watched and to keep hidden all evidence of the nurseries of silkworms and mulberry trees. The process of making silk was guarded in China as carefully as nations today guard their atomic secrets. The Chinese kept the secret of silk until well after the birth of Christ.

Then, in the second century A.D., according to tradition, the secret was carried to Korea as part of the Buddhist religion. From Korea, silkworms and mulberry trees spread to Japan. But, even after the Japanese knew that silk came from the cocoons of the silk moth, they were not able to make silk cloth until they were taught the things the Chinese had learned. It is said that four Chinese girls were captured and carried to Japan to teach the Japanese. As you will see, in every land where silk was to be produced, it was necessary for someone first to teach the people how to hatch the eggs, to feed and care for the growing silkworms, to kill the pupae before they break through the cocoons, and then to unwind the silk. So greatly did the Japanese appreciate the teachings of the Chinese girls that temples were built to honor them in Japan. By the time of the twenty-first Mikado in Japan (459-479 A.D.), farms were planted with mulberry trees and the raising of silkworms was a thriving industry. From that time to this, Japan has been a leader in producing silk.

India, too, learned about sericulture at an early date. A prince of India took as his wife a Chinese princess. When she journeyed from her homeland with her caravan of gifts and silken wedding garments, she carried with her some eggs of the silk moth. She had them hidden in her headdress. When she arrived in her new home, she taught her husband's people how to raise silkworms and feed them on the leaves of a kind of mulberry tree that grew wild in India. Then she taught people how to unwind the cocoons and weave the silk into cloth. From India, the secret of silk spread to Persia and to other countries of central Asia.

To those in the western lands, however, silk was still a mystery. They bought silk cloth from Persia, India, China, and other eastern lands, but they had no idea how to make any of their own.

All learned men knew what silk was, but none knew where it came from or how it was made. Many of them wrote of silk and gave their own ideas about it. Aristotle, the great wise man of ancient Greece, who lived more than three hundred years before the birth of Christ, was one of the first of the ancient writers who came somewhat close to the truth. In his book, *The History of Animals,* he wrote that silk was made by some kind of insect. Other writers of his period and later had interesting explanations. Here are some of the ways in which they explained the origin of silk:

> It comes from the colored petals of flowers that grow on the deserts of China.
>
> It is combed fleece from silken leaves of a tree.
>
> It is gathered from the boughs of trees.
>
> It comes from an animal that is twice the size of a beetle. The animal is something like a spider. It eats greens until it bursts open. Silk thread, already spun, is found inside its body.
>
> It is made from some kind of soil that is marvelously soft. The soil is sprinkled and then combed. From it comes silk cloth.
>
> It is made by a worm that spreads a web, which floats on the air like a spider web. The worm eats its web and is killed. The silken threads can then be unrolled from the inside of the worm's stomach.

These explanations sound fantastic today. And yet, when you raise some silkworms of your own, you will wonder if these ancient ideas are any more fantastic than the true story of silk.

3:
The Silkworm and Its Relatives

First of all, a silkworm is not really a worm at all. It is the wormlike stage in the life of an insect, the silk moth. The silk moth, like other animals, has a scientific name that describes it. Its name is *Bombyx mori,* which means mulberry silkworm.

All insects, as you probably know, go through different stages. Ants, bees, grasshoppers, mosquitoes, butterflies, and moths are a few of the insects you know. In their grown-up, or adult, stage, they have certain things in common. They all have feelers, or antennae. They all have six legs. They all have bodies that are divided into three parts – head, thorax, and abdomen. There are over 500,000 kinds of insects in the world. All of them have these characteristics when they are adults.

Often their adult life is a very short one. And in other stages, they look completely different. How different a caterpillar is from a butterfly! Yet it is the same insect, but in a different stage. And how different a silkworm is from the adult silk moth it will become.

Every insect begins life as an egg. Then, depending on the

kind of insect, it goes through two or three changes. Changing from one stage to another is called metamorphosis, which means changing form. Insects that change three times go through the following four stages.

First, there is the *egg*. From the egg comes a *larva*, which is a wormlike creature. The larva eats and grows. At the end of its growing time, it usually makes a cradle of some sort and goes to sleep. For the silk moth and many other insects, the cradle is a cocoon made of silken threads. During its sleeping stage, the insect is called a *pupa*. While the pupa sleeps, its body changes in a marvelous way. Its fat, wormlike form becomes thinner. Six slender, jointed legs grow. And most wonderful of all are the wings, usually two pairs, that change the crawling insect into a flying one. When the pupa stage is over, the insect works its way out of its cocoon. It opens its damp, crumpled wings and dries them in the air. When the wings are dry, the *adult* insect flies off. In time it finds a mate, and the female lays eggs. Then the adult insects die. The eggs hatch into new larvae, and the life cycle goes on, over and over again.

Bombyx mori belongs to the large group of insects that have a four-stage metamorphosis. As you raise your own silk moths, you will see them going through these four separate stages: egg, larva, pupa, and adult. It is in the larva, or caterpillar, stage that the *Bombyx mori* may rightly be called a silkworm. Then it is slow, fat, and crawling.

In the adult stage, the silk moth has antennae like thick little feathers. Its body is in three separate parts: head, thorax, abdomen. It has six jointed legs and two pairs of wings. But it can fly only a few feet at the most. Some cannot fly at all. This is because silk moths have been bred for centuries for one main purpose – to make silk. Generation after generation, the fattest, heaviest, and best silk spinners have been picked out and their eggs used for new silkworms. Naturally, these fat, heavy moths were not the best ones for flying. And so, as the silk moth became a better and better silk maker, it became a poorer and poorer flier. Now the silk moth has wings that are tiny and frail for its thick, heavy body. No wonder it can hardly get off the ground! But the silk moth no longer needs to fly. It is now taken care of, and everything it needs is nearby.

In most ways, *Bombyx mori* is like the insect group to which it belongs. All moths and butterflies are classed together as *Lepidoptera*. The word, Lepidoptera, means "scaly

Female silk moth laying eggs. *(Japan Silk Association, Inc.,* and *Japan Raw Silk Exporters Association)*

winged." The lovely colors on the wings of moths and butterflies are from tiny scales that lie in even rows, overlapping like shingles on a roof. Whenever the wings brush against something, some of the scales come off. Have you ever touched the wing of a moth or butterfly? If you have, you probably found bits of fine, powdery dust on your fingers. Under a microscope, the dust shows up as delicate little scales of different colors.

When the tiny scales are scraped off both sides of a wing, the thin wing covering can be seen. It is a kind of colorless, transparent skin, looking something like cellophane. It is stretched between the wing veins, which support the wings and also carry the insect's blood. This is not red blood like ours but yellow or green blood, depending on the kind of insect.

The scales on the wings of butterflies are often brightly colored. They give the butterflies their brilliant beauty. The colors and patterns help conceal butterflies from their enemies as they fly from flower to flower in the sunlight. Moths, however, are usually less brilliantly colored than butterflies. Many of them are white, gray, tan, or dull, mottled mixtures. Their pale colors and shadowy patterns help to conceal them as they fly in the dark of night and in the twilight and early morning hours.

Difference in color and brightness, then, seems to be one way of telling moths from butterflies. Usually, though not always, butterflies are brighter than moths. However, it is not easy to tell at a glance which are moths and which are butterflies. There are a number of other differences that will help you to tell one from the other. The feelers, or antennae, of a butterfly are like two fine wires with dots on the ends. Those of a moth are usually like little feathers. *Bombyx mori,* you remember, has thick, feathery feelers.

Butterflies have bodies that are usually thinner and longer than those of moths. When at rest, butterflies fold back their wings. Moths rest with their wings spread out like the wings of an airplane. No doubt you have often seen dainty butterflies with their brilliant wings pertly erect as they paused on the petals of flowers. Or at dusk, perhaps, you have found the shadowy forms of moths resting with outspread wings on a night-blooming flower.

Perhaps, too, you have found cocoons in the bushes and trees that grow near your home. You may have collected cocoons and kept them over the winter and watched the adult moths come out in the spring.

Many moths spend the whole winter in the cocoon stage. The adults are not ready to come out until the warm days of spring. Then the mating takes place, the eggs are laid, and, after weeks, the larvae, or caterpillars, hatch out. They eat and eat and grow and grow. Then, usually in the fall, they spin their cocoons and change into pupae, ready for their winter-long sleep. In order to see the complete life cycle of most moths, you would have to follow one for a whole year. Some kinds of moths take even more than a year to live through all four stages.

Bombyx mori, however, can do it all in a few weeks! Instead of living through the winter as a pupa, wrapped in a cocoon, the silk moth spends the winter as an egg. The egg hatches out in the spring, and the insect goes from one stage to another right before your eyes – egg, larva, pupa, adult – all in about two months.

So, before you start out to be a silk-moth farmer, you will need to find out about mulberry trees. You will want to know one when you see one. Probably there are some near your home or school, for mulberry trees now grow in many places all over the world.

4:
The Mulberry Family

It is hard to say why silkworms feed only on mulberry leaves. Most animals eat a nibble of this, a taste of that, and a bit of something else for dessert. Not so, the silkworm. It eats the same thing, morning, noon, and night – the leaves of the white mulberry tree.

Other kinds of leaves have the same chemicals as those of the white mulberry. But it seems that only this one kind has exactly the right amounts of all the things that the silkworm needs. Somehow, the chemistry in the little body of *Bombyx mori* is able to change the raw materials it gets from mulberry leaves into a liquid that becomes silk. Other insects can and do spin silk. But only the mulberry-eating *Bombyx mori* can make the kind of silk that is exactly right for weaving into silk cloth. Wherever silkworms are raised, mulberry trees must be grown.

Originally, mulberry trees grew wild in China and Japan. As sericulture became more and more important, laws were made ordering all farmers to plant mulberry trees in order to feed the silkworms.

Today in Japan, where great numbers of silkworms are raised, there are large farms or plantations of mulberry trees. Girls used to climb the tall trees in order to pick leaves for the silkworms. But now, Japanese plant scientists have developed a kind of white mulberry tree that grows near the ground like a bush. Each winter the dwarf trees are cut down to low stumps. New branches, with fresh new leaves, grow out in the spring. The leaves are always near the ground and easy to pick. The girls cut the leaves with scissors or with small knife blades that they wear fastened to their fingers.

The easiest way to get leaves for your silkworms is to find a white mulberry tree in your neighborhood. One big old tree will supply more than enough leaves for many young silk-moth farmers. It is said that mulberry trees are the greatest leaf producers of all plants. This is good, because, as you will find out, silkworms are great leaf eaters.

When you begin a mulberry hunt in your neighborhood, tell all of your friends and their parents what you are looking for. Perhaps you can write to your local newspaper about your search. A little news story in the paper will reach many people. Anyone who has a mulberry tree should not mind your getting leaves from it every day or two. It will not hurt the tree in any way. You will learn how to collect and take care of the leaves as you learn more about raising silkworms.

To get ready for a mulberry-tree hunt, you will need to find out what a mulberry tree looks like. And you may want to find out something about the mulberry family in general. It is a large family, including some other plants that you may know – the Osage orange, fig, hemp, rubber, and breadfruit.

Trees, like all plants and animals, have scientific names as

Mulberry leaves. *(Japan Silk Association, Inc.,* and *Japan Raw Silk Exporters Association)*

well as common names. All mulberry trees are called *Morus,* written with a capital letter, followed by a word that describes its particular kind. *Morus alba,* for instance, is the name of the famous white mulberry, favorite food of silkworms. You can now see where *Bombyx mori* got its second name. The word *mori* refers to *Morus,* the kind of tree on which it feeds.

All mulberry trees have the same general characteristics. The sap, the liquid that is inside the trunk and branches, is milky. It feels sticky on your fingers. The bark is scaly. The branches are slender and round. The leaves do not grow directly opposite each other, but alternate on the twigs. The leaves have uneven edges, sometimes with deep cuts or scallops. Two leaves, even two on the same branch, may be different in shape. One may have deep scallops, or lobes, and the next one may have no lobes at all.

The leaves of mulberry trees stay green through the late fall. Then they turn brown and fall off. Late in winter, new little leaf buds appear on the bare branches. In spring the buds grow and open into tender young leaves. By summer, the leaves have become three to seven inches long and a deep, shiny green.

The flowers come with the leaves. Mulberry trees have two kinds of flowers, male and female, usually on separate trees. The male flowers form short, spiky catkins. The female flowers form catkins that are longer and thinner. Pollen is blown or carried by insects from male to female flowers. Bits of pollen grow down into the seed-making parts of the female flower and make the seeds grow and ripen. Each seed grows inside a tiny globe of fruit, and groups of these tiny fruits make clusters of little berries. The fruits of mulberry trees are not important foods. They are important only for the seeds that are in them because the seeds can grow into new mulberry trees.

There are at least twelve kinds of mulberry trees growing in different parts of the world. Here are facts about some of the most common.

The white mulberry tree (*Morus alba*) probably grew first in China. During the past two thousand years it has spread over Asia, Europe, parts of Africa, and North America. The trees did not spread naturally but were grown by people who were raising silkworms. The white mulberry is a tree with a rounded top. In its natural state, it grows to be from twenty to forty feet tall, with spreading branches thickly covered with leaves. Its leaves are smooth above and below. Its fruits are small and colored a pale pinkish white. The white mulberry makes a good shade tree. The only reason that it is not more popular is that its berries make a sticky mess when they fall from the tree.

The red mulberry tree (*Morus rubra*) grows naturally in the United States – or, as a scientist would say, it is native to this country. It has been growing well here since before white men came to America. The fruits look something like blackberries when they are ripe. They are larger than the fruits of *Morus alba*. Mulberries are occasionally eaten by people and can be made into jelly, although they are not very sweet and do not have much taste. Birds like them, however, and certain songbirds choose them as their favorite fruits.

The black mulberry tree (*Morus nigra*) is not native to the United States, though it now grows well here. In form and fruit, it is similar to the red mulberry. *Morus nigra* once grew naturally in Greece. When silkworms were first brought into Europe, they were fed leaves from black mulberry trees until the more desirable white mulberry trees could be grown.

The paper mulberry tree is commonly grown today as an ornamental tree in our southern states. It differs from other mulberry trees because of its unusual bark and its large, rough, hairy leaves. In ancient times, the paper mulberry tree was used in China for making paper. The outer bark was stripped off the tree. The inner bark, which grows next to the wood of the tree, was taken off. It was boiled and pounded until it became a soft pulp. The pulp was then made into paper. Today there are Chinese who still make mulberry paper in exactly the same way.

On a diet of white mulberry leaves, silkworms spin their best silk. But if a white mulberry tree is not available, you can feed your silkworms with leaves from a red or a black mulberry tree. Silkworms have been known to eat leaves from the paper mulberry, too, but only young, tender leaves. Silkworms seem to need some kind of mulberry leaves in order to thrive.

There is one exception, and this may help you if you live where there are no mulberry trees at all. There is one other plant of the mulberry family – a distant relative – that the silkworms will eat. This is the Osage orange, a tree or shrub that may be found almost anywhere in the United States. It was once widely used to make living, growing fences around farms in the West and Middle West. It grows from fifteen to fifty feet tall. You can easily recognize an Osage orange tree by its green, orange-like fruits. These fruits look and even smell like green oranges. But inside they are dry and pulpy. The leaves are long, pointed, and very shiny. The twigs have thorns on them. The inner bark is bright orange in color.

If you have no luck in finding a white mulberry tree, look for some Osage orange trees. Though their leaves look quite different from those of the *Morus* group, they will give your silkworms what they need in order to grow, develop, and spin silk.

If you are not able to find either a mulberry tree or an Osage orange in your neighborhood, you can still raise silkworms. But first, you will have to grow some trees of your own. There are three ways of getting started.

You can begin, as some of the early explorers did, with

some seeds. Go to a nursery or to a tree farm and get some seeds of the white mulberry tree.

Plant the seeds in flowerpots that are filled with good rich soil. Keep the soil damp. The seeds will sprout and grow into baby trees. When the plants are a few inches tall, pull out all but one plant from each pot. Keep the young trees growing in the flowerpots until they are as tall as the pots are. Then loosen the soil and take out the plant without disturbing the roots. Plant it in a place where you have dug a nice soft bed of earth. It should have good soil, sun, and plenty of water. In most places in the United States spring is the best time to start a new tree.

Another way to start your own mulberry trees is to root some twigs. The best time to do this is in the spring or early summer. Cut a few growing ends of twigs from a mulberry tree. Cut each one just below a leaf bud. Sometimes tree twigs will root best in sand or soil, sometimes in water, and sometimes they will not root at all. Try a number of them and hope that at least one will grow roots. Try some in sand or soil in a flowerpot. Just stick the twigs in, water the pot well, and make sure that it never dries out. Keep it in a shady place that is warm but not hot. If roots form, the twig will begin to grow and the leaf buds on the twig will open. When this happens, plant the rooted twig outside in good soil.

You may also want to try to root some twigs in water. Just put the bottoms of the fresh-cut twigs into a jar of water, stand the jar in a shady spot, and keep adding more water as the water evaporates. You will know when roots begin to form because you can see them growing out from the side of the twig near the bottom. When the roots are fairly long and the twig has the look of a growing plant, it is ready for planting outdoors in good soil. Your rooted twigs will need the same care as the young seedling trees.

The easiest and surest way to get a tree is to buy a young one growing in a one-gallon or five-gallon can. You can buy young trees from a nursery or a tree farm. Mulberry trees are not expensive. The man who sells you the tree will tell you how and where to plant it so that it will thrive in your particular climate. Mulberry trees grow well where they have good soil, plenty of water, and good drainage.

In some schools, classes of older children raise silkworms every year as science projects. They observe and make records of the complete metamorphosis of the insects. Younger children raise silkworms as pets or to get silk from the cocoons. Where there are no mulberry trees already growing, parents in the P.T.A. sometimes buy mulberry trees and give them to the schools. The trees are planted in schoolyards, where they become beautiful shade trees, as well as supply food each spring for silkworms.

It would be helpful, wouldn't it, if silkworms would learn to thrive on lettuce or grass or spinach or celery tops? Sometimes they *will* eat these things if they get very hungry. And they may keep alive by eating them and even grow for a while. But they will probably not live through the complete life cycle, and they will certainly not be able to spin good silk unless they have mulberry leaves to eat. As you raise your own silkworms, you may enjoy experimenting with different foods to find out what happens to silkworms that are *not* fed on mulberry leaves.

5:
Begin with the Eggs

The egg of the *Bombyx mori* is very tiny. It has been likened to a poppy seed, to the head of a pin, and to a period at the end of a printed sentence. It takes 40,000 of them to make a single ounce!

In Japan, each farmer used to raise all of his own eggs. He washed the eggs in clear cool water. The fertile eggs, those that will hatch into silkworms, would sink to the bottom. The infertile eggs, those that would not hatch, floated to the top, where they were skimmed off.

Today, you can buy fertile eggs from one of the scientific supply houses listed on page 154. You will find that these companies sell eggs in batches of a hundred. This is enough to start several silkworm projects.

If you know someone who raises silkworms, you can probably get a few dozen eggs free. If so, you will want to do as the ancient silk farmers did, test your eggs to find out which ones are fertile.

You can get your eggs any time of the year. But you must

keep them cold and dry until it is time to hatch them. This means until the time when the tiny leaf buds are swelling and showing green on the mulberry trees.

To keep the eggs from hatching before their food is ready for them, put the eggs in an envelope and put the envelope into a jar or box with a cover. Then put the whole thing in the refrigerator and keep it there until the first leaf buds look fat and green. (Note: Do not put the eggs into the freezing compartment nor into a deep freezer. You want the eggs to be cold, but not frozen.)

In China and Japan, the silk farmers used to bury the eggs in the ground to keep them cold during the winter. But now they keep them in refrigerators just as you can do.

Watch the mulberry trees, and as soon as you discover the first leaf buds, take out the eggs and leave them in a warm room. The eggs will hatch at the same temperature that you need for comfort. Do not put them in the sun. Scatter the tiny eggs on the bottom of a cardboard box. Shoe boxes and candy boxes make good trays for silkworm rearing.

Put a couple of eggs on a piece of white paper and study them through a magnifying glass. If you can, look at some through a microscope. You will find that they look like tiny beach pebbles. When first laid, they were light yellow. But now they have the colors and textures of smooth dark stones — gray, reddish brown, and dark green. They are oval in shape, with what seems to be a dish-shaped depression in the middle. Study the eggs closely every few days. If you have a way of magnifying them twenty or thirty times actual size, you will be able to watch the tiny larvae growing and developing inside the eggs.

When you stop to think about it, these tiny eggs are really quite marvelous. Each little egg has in it not only the live beginning of a new animal, but also enough liquid food to keep it alive until hatching time. The eggs will not change as long as they are kept in a cold place, not even if you keep them cold for several years. But just as soon as they are allowed to get warm, they begin to change. Life has been in every little egg all the while. It needed only warmth to start it growing into a new animal. Even through an ordinary magnifying glass, you will be able to see some changes in the eggs. They suddenly seem to *look* alive after they have been in a warm room for a few days. The outer edges get darker, and there seems to be an empty space in the middle.

After the eggs have been in a warm place for about a week, study them very carefully. They may be hatching out any time now. Seven to ten days is the usual incubation period. At this stage, if you look at an egg through a powerful magnifying glass, you can see the baby larva. The dark body is curled around the outer rim of the egg. You can see tiny black dots, which look like eye spots but which are really little feelers on the larva's head. The depression in the middle of the egg looks white and empty by now. When your eggs look like this, you should have your mulberry leaves ready. These baby silkworms come out of their eggs with a built-in appetite. Locate the leaves but do not cut them until the first silkworm has crawled out of its shell.

For generations, silk farmers in China and Japan had an unusual way of keeping the eggs warm at hatching time. They would wrap them in cloth or paper and wear them next to their skin, or they would put the eggs into little cotton bags

Newly hatched silkworms. *(Japan Silk Association, Inc., and Japan Raw Silk Exporters Association)*

and sew them into the padded clothing of the children. There are still many villages where the silk farmers hatch their eggs in the old ways, using the heat of the human body.

The hatching of the baby silkworms is interesting to watch. After the first one breaks out of its egg, others will follow. It may take three or four days for all of the silkworms in a single batch to hatch out. As soon as you notice the first tiny dark creature dragging its empty white egg shell across the bottom of the hatching box, watch through a magnifying glass. You will see some of the insects freeing themselves from their eggs.

As a tiny silkworm pokes its shiny black head out of its egg, its jaws are working furiously. It actually bites its way out.

As soon as it has bitten a hole in the shell, it pushes out its large head. Then the front part of its body comes out. It helps itself out by means of the three pairs of tiny pointed legs that grow from the part of its body directly behind the head. Spaced along the rear three-fourths of its body are five pairs of false legs. They look like thick, stumpy legs, but they are not true legs. They are called prolegs. They support the silkworm's body as it moves. Watch the way your silkworms use their prolegs. They often seem to be serving as the animal's true legs. Its small, jointed real legs are used for climbing and holding rather than for walking.

The newly hatched silkworm is from one-thirty-second to one-sixteenth of an inch long. Without a magnifying glass you will see it only as a tiny, thin line, no longer than a comma on this page. But through a glass, you will see an animal with interesting and unusual features. Its head is its largest part. It has to be large to hold its powerful jaws, which from the beginning are its busiest parts. The head looks black and shiny as though it were made of patent leather. The body is brown, mottled with light colored spots. All along its body are tufts of long, stiff-looking black hairs. The Japanese call a newly hatched silkworm *kego*, which means "hairy baby." Silkworms have these hairs only in the newly hatched stage. The hairs keep each silkworm protected from other silkworms. In other words, every little silkworm has to keep its distance. Without the stiff hairs on their bodies, the tiny creatures might crowd together and smother each other.

It takes a larva only a few seconds to bite its way through its shell and to drag the front part of its body out. Then, for many minutes, it may move about with more than half of its

Silkworm emerging from egg. *(Japan Silk Association, Inc.,* and *Japan Raw Silk Exporters Association)*

body still encased in the egg. It crawls around on the front part of its body, dragging behind its swaying shell like a fat white bustle. Little by little it works the rest of its body out of the egg. The shell is now completely white and looks like a tiny version of the shell from a hen's egg. Only the hole through which the larva emerged looks different. All around the edges of the hole are dark brown marks as though the white shell had been charred by fire. Something from the jaws of the silkworm seems to put these dark stains on the white egg shells. Though the empty shells *look* like tiny hen's eggs, they are not nearly so brittle. Try to crush one with your fingernail. You will find that the shell is quite tough. It can

Finely cut leaves for baby silkworms. *(Japan Silk Association, Inc.,* and *Japan Raw Silk Exporters Association)*

stand a great deal of bumping and banging without breaking.

After the tiny larva has its whole body out of the egg, it still drags its shell around. There is a thin strand of silk that reaches from its body to the inside of the egg.

The larva is very active. It keeps moving until it finds what it is looking for – mulberry leaves! For their first feedings, silkworms need tender, young leaves. If the mulberry leaves are not quite out when your larvae begin to hatch, collect some of the leaf buds that grow along the twigs and branches. Slice the buds open. The baby silkworms can eat leaf buds for a few days until some leaves come out.

If the leaves or leaf buds are wet from rain or dew, dry

them off. Silkworms may die if they eat wet leaves. They get all the water they need from the plant juices inside the leaves. To feed your baby silkworms, use a knife to chop up the crumpled green leaf material from the inside of leaf buds or chop up some small, tender leaves. In either case, chop the leaf material until it is very fine. (As the silkworms get older, you can feed them larger pieces.) Lay the pieces on some netting like cheesecloth. Then put the netting right on top of the eggs and the hatching silkworms. Watch what happens.

As each larva gets out of its egg, it moves toward the green leaves. It crawls easily through the fine-mesh holes in the netting. But its egg shell is usually left behind. So, by placing the leaves on cheesecloth netting, you quickly separate the free larvae from the unhatched eggs and the hatching larvae. As each tiny silkworm frees itself from its egg, it climbs up to the top side of the netting and finds a place on a bit of green leaf.

If you do not have cheesecloth or any other kind of netting, you can lay the cut-up leaves on a sheet of paper in which you have punched many small holes. Then lay the paper right on top of the hatching larvae. The baby silkworms will crawl up through the holes in the paper to get to the green leaves.

As soon as you have about a dozen of the larvae on the mulberry leaves, lift the leaves, larvae and all, and put them gently into a new box. Then put more leaves into the hatching box, and when these have collected a dozen more of the tiny feeders, put this next group into its own box. From twelve to twenty silkworms makes a handy number to care for in a single box.

The tiny, new-hatched larvae are far too delicate to be

Silkworms feeding on mulberry leaves. *(Japan Silk Association, Inc., and Japan Raw Silk Exporters Association)*

picked up with the fingers. The traditional way of moving silkworms is to let them crawl onto a leaf and then move the leaf.

You will find that your silkworms will stay right inside their boxes. This is one reason why they are easy to raise at home or at school. You need no cages, no covers, no screens – just ordinary cardboard boxes from shoes or candy. In fact, silkworms can be kept with no boxes at all. They can be raised on pieces of newspaper, and they will stay right on the papers.

Silkworms have been raised and cared for by man for so many thousands of years that they are now completely dependent. Even if they are starving to death, silkworms will not crawl off in search of food. They will just stay in their boxes or on their trays or papers until they are fed – or until they die from lack of food.

If modern silkworms are turned loose, they cannot care for themselves. Thousands of years ago the wild moths lived natural lives on the branches of wild mulberry trees, but because they have been tamed, or domesticated, for so long, they can no longer survive without care and protection from man.

And so, if your newly hatched hairy babies are to grow into big, fat spinners of silk, you will have to watch over them and feed them and care for them every single day for at least a month.

6: Eating and Growing—the Larva Stage

As your silkworms grow, you will see their bodies change. Watch them at first through a magnifying glass. Later you will be able to see them clearly without the aid of a glass.

In most ways, the silkworm's body is like the bodies of other kinds of caterpillars. It has thirteen parts, or segments. They fit together like the sections of an accordion. As the worm moves, its segments spread out and then squeeze together, spread out and squeeze together, over and over again.

There is a pair of legs on each of the three segments just back of the head. These tiny legs are often raised with the head while the animal is feeding. On the sixth, seventh, eighth, ninth, and twelfth segments there are pairs of fat stumps – the prolegs – that look like legs but are not true legs. When you see a silkworm inching along on its stumpy bumps, you may think it has ten legs in back and six legs in front. In fact, you may even read that caterpillars have sixteen legs. But the

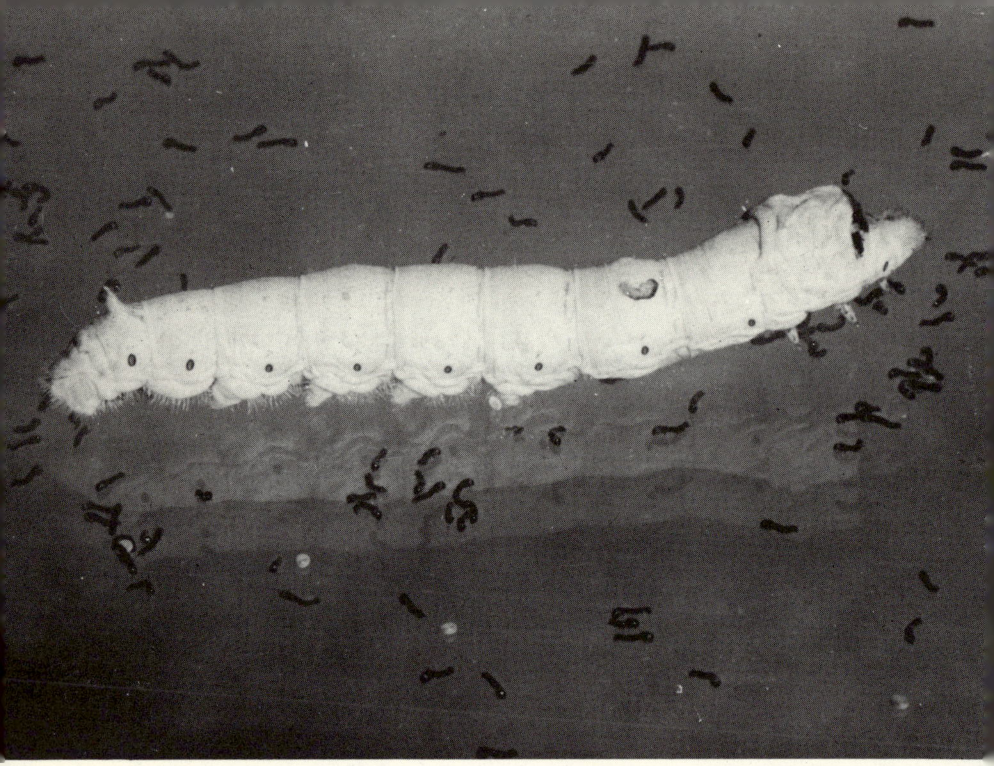

A full-grown silkworm with tiny newly hatched ones. *(Japan Silk Association, Inc.,* and *Japan Raw Silk Exporters Association)*

truth is that only the tiny jointed ones on the front three segments are true insect legs. The other ten are just extensions of the segments themselves.

The head of the silkworm is mostly mouth. The jaws, or mandibles, are strong and efficient. They work like scissors in biting into the mulberry leaves. Teeth in the mandibles chew up the leaves. The silkworm's eyes are weak, and they probably can see very little. They seem to find their way by using their palpi, which are little organs of touch located on their lips.

In appearance, there are usually two kinds of silkworms. One kind has many little dark spots on its body. The other kind has plain, clear skin. The Japanese think that the plain ones are more beautiful. They call them princess silkworms.

Like all insects, silkworms breathe very fast. They have no real noses and do their breathing through little holes, called spiracles, in the sides of their bodies. You can easily find the spiracles on any grown silkworm. A silkworm has no lungs, no heart, and not even a brain. Its strange green blood moves in tubes through its body. Its air supply also moves in tubes. There are many nerves that keep the blood and the air moving. In most ways, the silkworm is like all other caterpillars. They are all big-jawed creatures without hearts, noses, or brains. And they are all enormous eaters.

When the silkworms are very tiny, they must be given fresh leaves from three to five times a day. Remember to give them

Kinds of silkworms. *(Japan Silk Association, Inc.,* and *Japan Raw Silk Exporters Association)*

only clean, dry leaves. They cannot eat leaves that are dusty or wet. If your leaves are not clean, wash and dry them carefully before you feed them to your silkworms. Chop the leaves up fine when you are feeding very small larvae. As the larvae grow, they can eat from larger and larger pieces. You will see that they chew around the outer edges of each piece. If you give them whole leaves or large pieces while they are still tiny larvae, the middle parts of the leaves will become dry and worthless before the silkworms get to them. So, by cutting up the leaves first, you can get the most food from your leaves.

You may want to do some experiments with different kinds of feeding. Put a few silkworms in different boxes. Feed those in one box only lettuce. Use lettuce that is slightly wilted, and be sure that there is no water on the leaves. Feed those in another box only the green leaves from dandelion plants. Try feeding those in a third box another kind of greens such as spinach or cabbage or the tender leaves from the tops of celery stalks. Give the rest of your silkworms their regular mulberry-leaf diet. (And remember, the leaves of the Osage orange may be used instead of mulberry leaves.)

Watch the silkworms that are eating different kinds of greens. You will probably find that some of the larvae will eat lettuce, celery tops, and other greens. But they will not grow as well as those that feed on mulberry leaves. And when the time comes for them to spin silk, they will probably not be able to do so. Their lives will end without completing the life cycle. Of course, you *may* discover some kind of food, other than the leaves of mulberry or Osage orange, that your worms can thrive on. If you do, your discovery could change the whole silk industry, for as far as we know now, the larvae of the silk moth are completely dependent upon the mulberry or its relative, the Osage orange.

The silkworm eats its own weight about fifty times during its eating and growing period. Eat, eat, eat! Morning, noon, and night the jaws work like pairs of scissors, cutting and tearing into the green leaves.

As it eats, the silkworm puts on weight very rapidly. Before its eating time is over, it increases in size more than 10,000 times! If a human baby, which weighs about eight pounds at birth, grew at the same rate as a silkworm, it would weigh more than 80,000 pounds by the time it had grown up! But there are few, if any, animals — and certainly no people — that eat and grow at the silkworm's fantastic rate.

While the rest of the silkworm grows very rapidly, its skin does not grow. And so, from time to time, the silkworm's skin gets too tight. When this happens, usually every six or seven days, the silkworm stops eating for about two days. It raises its head in the air and moves it from side to side as though it feels annoyed at something. This is the beginning of the molting time or what the Japanese silk farmers call "the sleeping days." The silkworm holds up its head all during its sleeping time.

But before it goes into its sleep, it binds its body to the floor of its box or tray. With its own silk threads it makes loops around its back and fastens the ends of the loops to the floor. These loops hold on to its tight outer skin, holding the silkworm in one place while it is in its strange sleep.

When the silkworm awakens, its old skin breaks open. Then, with the old skin held down by the silk threads, the silkworm slowly wriggles its body out. It is covered with a new, larger

Silkworm molting. *(Japan Silk Association, Inc.,* and *Japan Raw Silk Exporters Association)*

skin. Now the animal can move easily once more. It is again interested in eating. Sometimes it breaks its fast by eating its cast-off skin.

Silkworms molt four times – on about the sixth, twelfth, eighteenth, and twenty-sixth days.

While the silkworm is molting, it is especially important to keep it in a quiet, somewhat dark place. At least do not let it stay long in bright light. The first and second moltings each take about twenty-two hours to complete. The third molting takes longer, about twenty-seven hours in all. And the fourth molting takes all of forty-one hours. As the silkworm gets longer and thicker, it takes more time to get out of its old, tight skin.

Silk farmers say that after the second molting a silkworm

is like a two-year-old child. At this stage, the animal is delicate and subject to sickness, especially indigestion.

Silkworms, like all other kinds of animals, need certain conditions in order to keep healthy and grow well. They need quiet. It is said that sudden loud noises, such as a violent thunderstorm, can interfere with their growth, so do not try to raise silkworms in any place except a quiet room. They also like plenty of fresh air but should not be left by an open window where wind or a draft can blow on them. They like a warm place, not hot, but a comfortable temperature between 65° and 75°. Most of all, they must be kept clean. Always wash your hands before touching them or handling their food.

The boxes in which you keep your silkworms must be cleaned at least once a day. This is very easy to do since the silkworms always stay right on the leaves. The droppings, the bits of digested food, fall to the bottom of the box. Have a few extra boxes and just lift the leaves, silkworms and all, from an old box and put them gently into a clean one. Then empty out the used box and it is ready for new tenants. From your relatively few silkworms, there will be no offensive smell at all — at least not if you clean the boxes once a day.

From time to time, as you move the silkworms to clean their boxes, you may need to separate them into more boxes to keep from crowding. Since the silkworms grow very rapidly, they will need more and more room as they grow.

One reason silkworms must be kept clean is that they can get a fatal disease. If one silkworm has the disease, it can spread to others if everything is not kept clean. The disease is caused by a germ that is passed from the moths to the eggs and then to the worms. The disease was known and feared for hundreds of years, but no one knew exactly what caused it nor how to fight it.

About a hundred years ago, about five million dollars' worth of silk was produced every year in France. And then, suddenly, the silkworms all over France got sick. It was an epidemic of silkworm disease. The sickness spread from France to Spain and then eastward to Turkey and China. For a while it seemed that the silk industry of the whole world might be wiped out by the disease that was killing the silkworms.

In France, the silk industry dropped to less than two million dollars' worth of silk a year. And it kept on dropping. Sericulture had been important in France for hundreds of years. The people who raised silk and those who wove it into cloth grew desperate. They asked their government to help save the silkworms.

The French government hired a great scientist to study the problem of the silkworm disease. His name was Louis Pasteur. You have probably heard of his work. It was he who discovered a way of preventing death for those who were bitten by mad dogs. It was he who discovered ways to kill germs by using heat. And so, when the French government asked Louis Pasteur to take over the disaster of the sick silkworms, they chose a scientist who was one of the most important of his time.

He worked on the problem for five years. During that time he was criticized by many people for being so slow. His critics did not realize that a scientist has to work carefully, step by step. There are many false starts. There are many failures. But a great scientist keeps working at a problem until he solves it – or until he convinces himself that the solution is hopeless.

The silkworm sickness was one of the problems that Louis Pasteur was able to solve. He found out that the silkworms were suffering from two kinds of infections, and he developed a way of preventing both. In doing so, he saved the silk industry of France and possibly of the world.

Today, where silkworms are grown for the silk industry, all moths are carefully examined. Any that show signs of having a silkworm disease are destroyed and their eggs with them. All other moths, eggs, and silkworms are kept in clean places in order to keep them free from disease.

Silkworms have enemies other than disease. Ants and mosquitoes can destroy a boxful of silkworms overnight. It is always a good idea to use an insect spray on the table or shelves where the silkworm boxes are kept. Also, spray around cracks in the rooms and around the edges of windows and doors where the harmful insects might get in. (Do not, of course, use the spray in or close to the boxes themselves. And *never* let any of the spray get on the silkworms or on the mulberry leaves.)

If you are raising silkworms in your classroom and if there are ants around, you may want to take special precautions to protect the silkworms during the night. Get some large cooky or cake pans. Put about a half inch of water on the bottom. Then put some rocks on the bottom of each pan. Do not let any rock touch the edge of the pan. Then stand the box of silkworms on top of the rocks, well above the water level. Do not let the bottom of the box get wet. The box will then be safely protected from ants and all crawling insects. The insects will not be able to get into the box because of the water. The best protection from mosquitoes and other flying pests is to keep the silkworms in a room that is screened or closed up during the night, and cover the boxes with mosquito netting.

All of this protection, cleanliness, and careful feeding is necessary if your little tiny new-hatched silkworms are to grow into fat, full-grown caterpillars ready to spin silk. More than any other kind of insect, silkworms depend upon constant loving care. This they need in order to get themselves ready for the job they do better than any other animal in the world – spinning silk.

7:
Spinning the Cocoon

People raise silkworms in order to use the cocoons for making silk cloth. But as far as the silkworms are concerned, the cocoon has nothing whatever to do with silk cloth. A silkworm does not obligingly spin a silk cocoon to make beautiful silks for people to wear. The cocoon serves a far more important purpose for the little animal that spins it.

Every caterpillar, whether it will be a moth or butterfly, makes some kind of shelter where it can quietly turn into an adult. In doing so, it is following an instinctive pattern of behavior that is probably millions of years old. Why does it do this? How does it know how? It has never seen a cocoon at the time it begins to spin its own. It has never watched another silkworm at its spinning. Yet, at the very first trial, the little animal is able to fashion a complete and beautiful thing. Each makes its own kind of cocoon. The silkworm's cocoon is never like the cocoon of a Cecropia, Sphinx, Promethea, or other kind of moth. It is always a silkworm cocoon and exactly like those spun by the millions and millions of other silkworms all

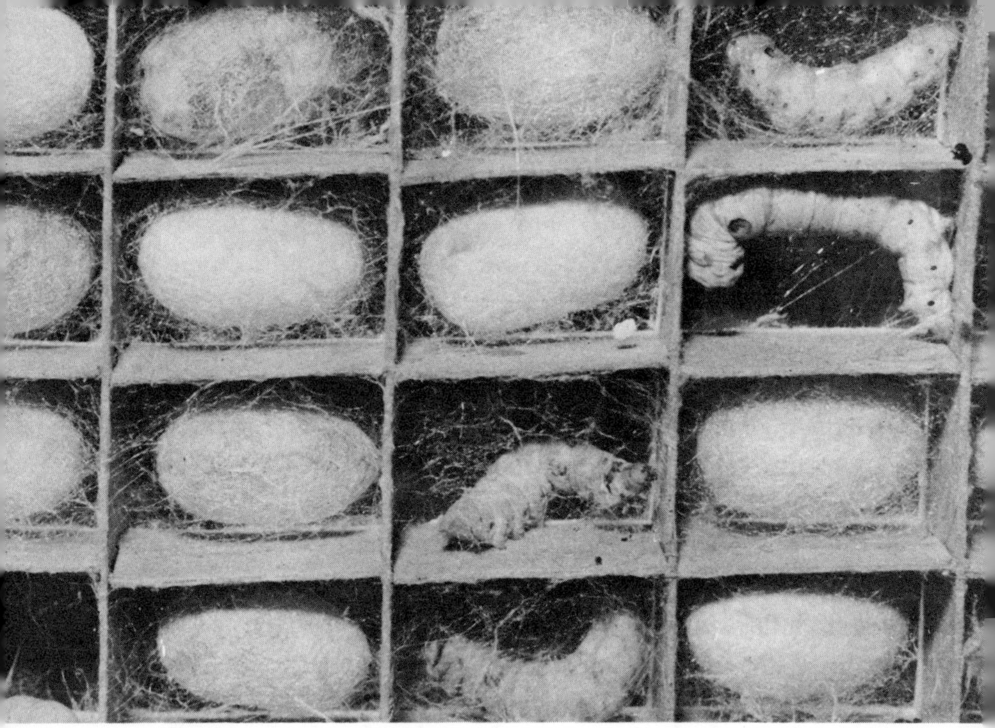

Spinning cocoons. *(Japan Silk Association, Inc.,* and *Japan Raw Silk Exporters Association)*

It stops eating. It lifts the front part of its body into the air and waves its head back and forth, back and forth. It seems to be nervous, restless, seeking. When your silkworms reach this stage, you will know that they are ready to spin their cocoons. They need no more mulberry leaves. The eating and growing days are at an end. What they need now are places on which to spin.

Gather pieces of dry twigs or the dry stems of weeds. Make them into little bundles shaped like Indian wigwams. Stand them in the corners and around the edges of the box. One by one, the ready silkworms will make their way to the supports you have made for them. Then each, with its marvelous built-in skill, will make the strong silken supports from which it will hang its cocoon.

If you live in a big city where it is hard to find dry twigs or weed stems, you can use an empty egg carton. Just lift each silkworm that is ready to spin and place it gently into one of the spaces in the carton. At the end of the spinning time each carton will contain twelve cocoons.

For thousands of years in China and Japan the silk farmers have provided little bunches of straws for their silkworms. The silkworms attached their cocoons inside the straw supports, which were then collected on trays. Today, silk raising, like everything else, has been made more modern. The modern Japanese silk farmer supplies square cardboard mounts that look like our egg cartons.

You may find it interesting to collect your cocoons in both the ancient and the modern ways. Use bunches of straws or twigs for some of your silkworms. Lift some of the rest into an egg carton, which is a good substitute for the modern Japanese cocoon mount. And you might like to try a third kind of support for cocoons. The third way is really the first and original way in terms of history. Just cut a small branch from a mulberry tree and bring it into the house. Put some of the ready silkworms onto the branch. See whether they will spin their cocoons as their wild, free ancestors did, on the branches and twigs of a mulberry tree.

As soon as you are sure that each silkworm has some place where it can hang its cocoon, settle down and spend as much of the next twenty-four hours as possible watching the spinning of the cocoons. Pick out a particular silkworm and observe everything it does.

First of all, it attaches itself to something, to the straws, twigs, branch, or to the sides of the cardboard carton. Its first silk is used to make a strong support for the cocoon. Then it bends its body into a U shape and begins to move its head in a figure eight. It spins the outside of its cocoon first. It builds it around its own curved shape in a series of figure-eight loops. It does not loop the figure eights over its body. It builds the wall of the cocoon by making rows and rows of tiny figure-eight loops, which gradually form a strong, solid shell. For each figure eight, the silkworm moves its head in a flat, circular path. It does this about once a second. Scientists estimate that it takes about 300,000 of these motions to complete a cocoon. Twenty-four hours after the silkworm has begun to spin, its body is hidden. It is entirely covered by the outer layer of the cocoon. Inside, the silkworm keeps right on spinning, day and night. Then, after three or four days of constant spinning, the silkworm stops. The silk-making liquid is all gone. The last tiny drop has been used up. The cocoon is finished. And there, within the cocoon, the silkworm molts for the fifth and final time. Its skin splits down the back, and the animal manages to work itself out of its old skin even though there is not room enough inside the cocoon for it to turn around.

After the silkworm has cast off the old skin, a new kind of covering begins to grow over its body. The covering is made of chitin, a kind of hard material like a very, very thin fingernail. Two or three days after the cocoon has been finished, the silkworm stage is ended. In place of the silkworm inside the cocoon there is a brown chiton-covered pupa. The name pupa

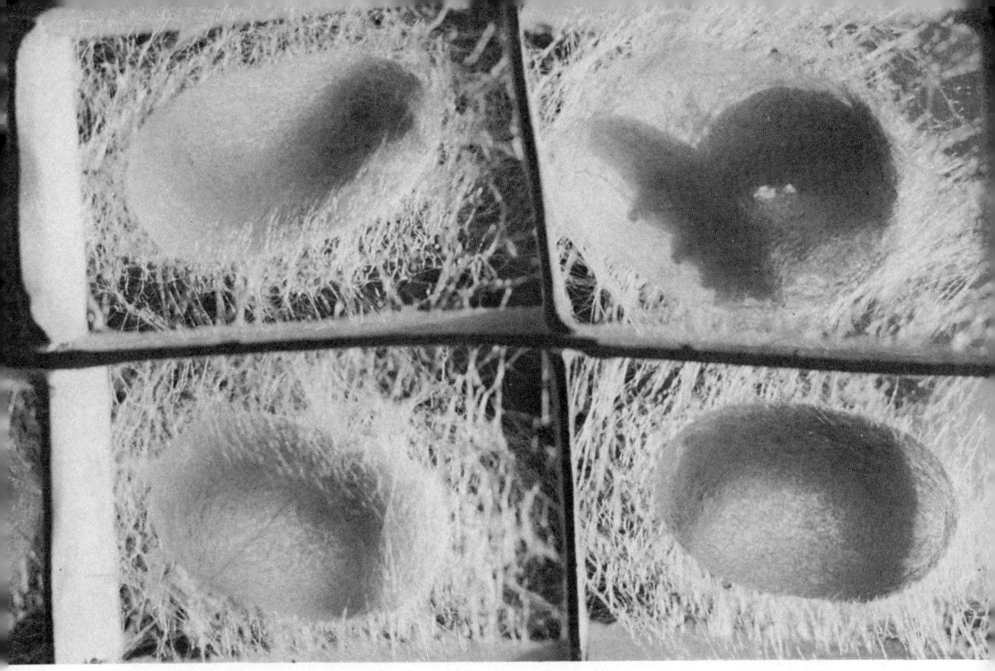

Close-up of cocoon spinning. *(Japan Silk Association, Inc.,* and *Japan Raw Silk Exporters Association)*

comes from an old Latin word for doll. It was once thought that insects in their cocoons look like little baby dolls wrapped in blankets. You may think so, too.

Cut open one of your cocoons. Use a single-edge razor blade or a very sharp knife. You will see the little brown pupa. It does look something like a little doll, doesn't it? You can save the open cocoon with its pupa for an exhibit showing the stages in the life of the silk moth, but first you will have to kill the pupa to keep it from changing further. You can do this most easily by putting it, cocoon and all, into an oven and heating it to 200° for a half hour. Once the pupa has been killed, you can keep it indefinitely.

The other pupae, however, cannot be kept, for, inside their

Cocoon cut open to show pupa. *(Japan Silk Association, Inc.,* and *Japan Raw Silk Exporters Association)*

dark and cozy cocoons, they are asleep but very much alive. And during each pupa's strange, secret sleeping, great changes are taking place.

This is nature's way with all moths. As caterpillars, they build their cocoons. They change into pupae and remain asleep, or dormant, while their bodies change. And then, usually when the warm days come in springtime, they break out of their cocoons as adult moths. But, in silk farming, only a few of the moths are allowed to do this. For the others, the life cycle is interrupted. It is brought to an end at the pupa stage in order to use the cocoons for silk. Only the moths that are to be used for breeding and laying eggs are allowed to emerge from the cocoons.

So, less than a week after your cocoons have been finished, you will have to make a decision. Do you want to collect the silk or do you want to see the life cycle completed and collect some new eggs? Perhaps it is most interesting to do both.

8:
Getting Silk from Cocoons

Within two or three days after the last cocoon has been finished, make up your mind which ones you want to use for silk and which ones you want to set aside. Those you set aside will be left alone until the adult moths break out of them. If you wait too long before deciding, all of the moths will emerge, leaving you no whole cocoons for reeling.

Take all of the cocoons out of their supports as soon as they are dry and hard – usually after three or four days. With your fingers, rub off the webs of coarse, tangled silk that have held the cocoons in place. Save the silk. You will use it later in an experiment.

Look at your cocoons. Feel the weight of them in your hands. Shake some of the cocoons gently. Experts can tell by the sound whether the pupa in a cocoon is dead or alive. Notice the shape, color, and texture of your cocoons. You will find that they are not all alike. Some are rounder than others. Some are shaped like fat peanuts. They may be pure white, cream-colored, or golden yellow. Chinese silkworms spin white

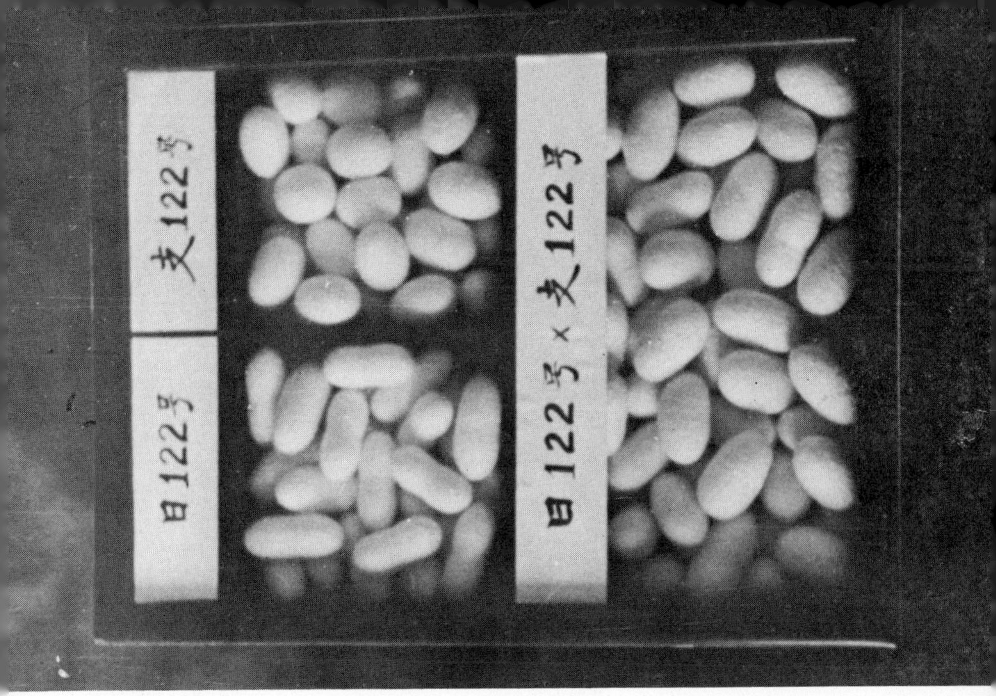

Kinds of cocoons. *(Japan Silk Association, Inc.,* and *Japan Raw Silk Exporters Association)*

cocoons. Those from Japan make cocoons that are creamy white or yellow. Italian cocoons are usually all yellow.

You will notice that some cocoons are bigger and heavier than others. An average cocoon weighs three-fourths of an ounce. Only a fourth of its weight is silk. The rest is the weight of the pupa inside. It is likely that the biggest, fattest, heaviest cocoons will have the most silk and also the largest pupae. You may want to do as the Japanese egg-producers do, keep some of the big cocoons for breeding purposes.

If so, pick out a few and string them on a thread. Use a needle and take a little stitch in the very outside layer of silk. Then hang the string of cocoons in a comfortably warm room. The cocoons should be on or near a plant or some tree branches so the moths will have something to climb on when they emerge. However, if this is not convenient, the moths will manage all right on a flat surface. It is not necessary, of course, to put your cocoons on a string at all. But by doing so, you can arrange them so that it will be easy to keep an eye on them later while the moths are getting out.

Now go back to the rest of your cocoons – the ones you are going to unwind for silk. Look at their rough, woolly surface. Examine them through a magnifying glass or through a microscope if possible. You will see a kind of kinky, gleaming fuzz all over each cocoon. You will see the twisty pattern of the figure eights in which the silkworm placed the strands around its body.

Pull off some of the loose, woolly silk that covers the outside of a cocoon. This silk is different from the single strand of fine silk that you will unwind later. The outside silk is in short, webby bits. It is coarser than the inside silk. It cannot be wound onto spools (or reels) because the pieces are too short and too uneven in weight. But it can be twisted or spun into silk yarn.

Now pick up the outside webs of silk that you had put aside and lay them in a pile with some of the loose silk from the outsides of your cocoons. Take some between your fingers and twist it. Pull it as you twist. When the strand gets too weak and thin, add more bits of the woolly silk. Then twist and pull again. Do you know what you are making? You are making a strand of spun silk. This is probably the first kind of silk yarn that was used.

In prehistoric times, in many places in the world, man learned how to spin different fibers into yarn, which was then woven into cloth. Fibers from the wool of sheep and other animals, fibers from the soft cotton bolls, fibers from flax and yucca plants – these were some of the materials from which early man spun yarn. The fibers were twisted together until they formed strong yarn or thread.

Probably the first silk cloth was made from spun silk. Perhaps, in prehistoric China, people learned how to pull apart the cocoons of wild silkworms and get the bits of silken fleece. The fleece could then be spun into coarse, shiny silk threads. The threads could then be woven on looms to make a heavy, bumpy silk cloth. This *may* have been the way the Chinese first made silk cloth. But we do not know. Spun silk is still made today. The good cocoons are no longer pulled apart to make it. The fleece for spun silk comes from the coarse bits on the outsides of cocoons and from cocoons that have been pierced or broken.

You have made a strand, perhaps a long strand, of spun silk. Now you can make a very, *very* long strand of the other kind of silk – reeled silk. The finest, softest, most shimmering silk fabrics are woven from threads of silk that are unwound from the cocoons – unwound yard after yard, just as the silkworm spun it.

The best time to begin the unwinding is eight or nine days after the cocoons have been finished. By then the cocoons will be completely dry and the chitin on the outside of the pupae will be hard. The first thing to do is to kill the pupae. This can be done without injuring the silk in any way.

Put the cocoons on a cooky sheet and keep them in the oven for a half hour. Keep the heat in the oven at about 200°. Or put the cocoons into a strainer and put the strainer into a pan that has boiling water on the bottom. Put a lid on the pan. Keep the water boiling until the steam has gone all around the cocoons. About fifteen minutes of steaming will kill the pupae. If the cocoons are boiled *in* the water, they sometimes get sticky and they usually begin to unwind themselves.

Once the pupae have been killed, the cocoons will keep indefinitely. Just spread them on papers until they are cool and dry. Then put them into a box and lay them away until you are ready to unwind the silk.

For thousands of years, silk farmers have killed the pupae with steam, boiling water, or dry heat. The process of killing the pupae is called stifling. Stifling is an important step in sericulture, modern and ancient. There is a record from very ancient times in China telling of stifling by packing the cocoons in salt in airtight jars. Another method used very long ago in China was simply spreading the cocoons out in the very hot sun. The rays of the sun cooked the living pupae to death. These ancient methods were slow and not always sure. And so steam or dry heat came to be preferred. Today the most modern method is radio high-frequency waves. These waves kill the pupae instantly without in any way destroying the silk.

Any time after the stifling you can begin to collect your own raw silk. This is perhaps the most exciting experience in the whole process of raising silkworms.

As you look at your little white or yellow cocoons, you will remember that cocoons like yours have supplied all the silk that has ever existed in the whole world. The embroidered robes of Chinese emperors, the brocades and velvets of kings and queens, the lustrous gowns and satin breeches of lords and ladies – all these came from delicate strands of silk unwound from cocoons. Silk, for hundreds and hundreds of years, was only for rulers and nobles. But today we all wear silk! Dresses, suits, ties, ribbons, underwear – there is enough silk for everyone, and every inch of it has been unwound from a little cocoon.

It is not hard to unwind a cocoon if you know how to do it. You will remember that the silkworm made its cocoon in little figure-eight loops. It built up the cocoon from these figure eights, working from the outside to the inside. Now you are going to take the cocoon apart. First, you are going to soften the silken loops. And then you are going to pull them off, inch by inch, from the outside to the inside of the cocoon.

Put about five cocoons into a pan of boiling water. Keep the water boiling for about five minutes. With a fork, keep turning the cocoons over and over. The hot water will soak off some of the sticky sericin, and the cocoons will begin to loosen up. Soon you will see some tangled threads around each cocoon. With a fork pick up the strands from one cocoon. Take the pan off the stove. Holding the loose strands in your fingers, dip one cocoon up and down in the water until you find a single thin strand of silk that pulls away easily. Put the end of that strand against the side of a round pencil. Turn the pencil in your fingers. You are now reeling a strand of silk.

The strand is very, very fine. You may have trouble keeping your eyes on it. If you do the reeling against a dark background, the whole thing is easier. Put some black paper or cloth on a table. Have another piece on the wall back of the table. Move the pan with the cocoons onto the black-covered table. Then you will be able to see the fine silken strand as it leaves the cocoon and stretches upward to your pencil. Of course, after you have turned and turned your pencil reel many times, the silk will show clearly on the reel.

How long do you think it will take you to unwind a single cocoon? Perhaps you can figure it out. One fat cocoon may have from 2,000 to 3,000 feet of silk in a single, unbroken thread. About a half mile of silk, more or less, is a long, long strand. Now, as you remember, it took the silkworm about four days to spin it. And the silkworm worked day and night, with no coffee breaks, no time off for meals, and not even time off to sleep. It worked twenty-four hours a day and as much as ninety-six hours on its cocoon.

In that time, if its strand of silk happened to be about 2,400 feet long, it spun at the rate of about twenty-five feet an hour – or about five inches of silk every single minute! You will probably find out that you cannot unwind a cocoon as fast as the little silkworm spun it, at least not on your slim pencil reel. You will wind and wind and wind. The kinky loops of silk will lift off the cocoon easily as it bobs up and down in the water. The kinks will straighten out, and smooth silk will roll onto your pencil. After you have been winding for a long, long time, look at your cocoon. It will seem as big as ever.

A bigger reel, of course, will do it faster. You may be able to make a hand reel by using a large spool and a dowel rod, supported in a cardboard box frame, or you can make a reel from a wood or plastic toy construction set. What you need is something on which to wind the silk, some sort of axle or handle with which to turn it, and a support of some kind for the whole thing to rest in. A large cardboard spool from

Christmas ribbon makes a good reel. Just slip it over a pencil and spin it around with your finger.

While you are unwinding your first cocoon, the silk in the other cocoons is also soft and ready to be unwound. One at a time, pick up the ends from each of the other cocoons and touch them to the strand you are winding. You will find that the strands from the separate cocoons will stick together. There is enough sticky sericin in the silk to bind the strands into a single heavy thread.

For thousands of years, cocoons were unwound by hand, using methods very like the ones you are using. The cocoons were first sorted. Those that were broken or stained or pierced by moths were put aside. They were later pulled into silk fleece for making spun silk.

The good cocoons were put into pans of boiling water. The water was kept hot by little charcoal fires. After the hot water had soaked off some of the sticky sericin, the strands of silk began to loosen up. The cocoons began to come apart. Girls would then reach into the boiling hot water to pick up the end of a strand of silk. Often their hands were burned in the hot water. The end of silk was put on a reel of some sort and the cocoon was unwound as it bobbed up and down in the hot water.

At first the reel was no more than a round twig that was turned in the fingers. Later, special reels were made for unwinding the cocoons. Some of the reels were worked by foot power. A girl used a foot pedal to make the reel turn. Other reels used water power and then steam power. The power made the reel turn, but the main work was still done by hand, usually by girls and women. In ancient China, it was said that one person, working from dawn to dark, could reel only a pound of silk in a week.

In Japan all reeling was done by hand until the 1930's, when the first silk-reeling machinery was imported from Italy and France. A place where silk cocoons are unwound is called a filature.

It is interesting to realize that no matter what kind of power turns the reels in a filature, the cocoons still have to be handled in the same old way. They still must soak and bob in pans of hot water. And the loosened ends of silk must still be caught up and lifted to the reel. Even in a modern filature, much of the work of unwinding the cocoons still depends on the skillful hands and watchful eyes of trained workers, usually girls.

The long strand that you are winding on your pencil, or on your homemade reel, is pure raw silk, exactly the same kind of silk that is collected in the silk filatures in Japan, China, Italy, and other countries that raise silkworms. If you can, look at a single strand of silk under a microscope. You will see the smooth, clear thread, or fiber, which gives silk its special beauty.

It will take you a long time to unwind, or reel, all of your cocoons. If a number of people take turns at the job, it will go faster. Eventually you will come to the end of each thread. You may find, at the end, some short bits of extra silk. And after the silk has been removed, you will find the dead pupa, the insect that was killed before it could change into a moth and break through the cocoon.

In olden times, silk farmers and their families used to save the pupae and use them as food. The little pupae are high in food value. Among the people who had little if any meat to eat, these bits of animal food were welcome. In more recent times, the pupae have been saved and returned to the soil. They have often been used to fertilize the mulberry trees.

In Japan during World War II there was once more a great shortage of food. Then, on many silk farms, families again used the silk-moth pupae for food. They cooked them in special ways and used them to improve their diet.

Today, four countries lead in the production of raw silk: Japan, China, U.S.S.R., and Italy. Other countries that are important silk growers today are India, Korea, Brazil, Turkey, Iran, Greece, France, and Spain. Japan, however, produces more than all the rest.

Thousands of tons of raw silk are produced every year. Just think, thousands of *tons* – and every ounce of it spun by little silkworms! And every inch of it was unwound from a silken cocoon.

9:
Completing the Life Cycle

While you have been busy reeling your long strands of silk, something important has been going on inside the larger, heavier cocoons. There, the insects have continued to develop and change. Each little insect, you will remember, ended its larva stage by shedding its skin for the fifth time inside the cocoon. Its body then grew a new kind of covering – chitin. It became a pupa. The pupa of a silk moth, like the pupa of any other insect, has one main job to do. It remains in a kind of sleep while it changes into an adult.

Under the hard chitin covering of the silk-moth pupa, wonderful changes are taking place. Two pairs of wings are forming. Six pairs of long, thin, jointed legs are growing out from its thorax. It is developing a pair of thick, feathery antennae. Its chewing mouth is disappearing. In its place grows a useless mouth, for there will be no more eating in the life of the silk moth. The long, thick, bare, wriggly body of the caterpillar changes into the shorter, more compact body of a winged insect. The wings, like those of all insects that belong to the

Adult moth emerging from cocoon (a large drop of solvent is still on its mouth). *(Japan Silk Association, Inc.,* and *Japan Raw Silk Exporters Association)*

Lepidoptera group, grow tiny scales, which give the wings their pearly white color. The body, too, is covered with scales, which give the body a furry look. Within ten days, there is no longer a sleeping, baby-doll pupa within the cocoon. In its place is a restless, full-grown moth, ready to push out of its silken prison and dry its crumpled wings in the sun and air.

Its way of getting out of its tightly wrapped case is most amazing. The baby silkworm, remember, was able to bite its way out of its egg, but the adult moth has no way of biting its way out of the cocoon. It has no teeth and, in fact, hardly any mouth, but it does have a simple and efficient way of freeing itself. It simply dissolves some of the silk! As a larva, its glands made silk. Now, as an adult, its glands make a chemical that dissolves silk.

With this homemade solvent, which comes out through its mouth, the moth makes an opening in one end of the cocoon. Then, bit by bit, it works itself out. This usually happens early in the morning, rarely after nine o'clock. One wonders what serves as an alarm clock for the emerging moth. What times its early-morning outbreak? Perhaps it is light. Perhaps in some marvelous but mysterious way the first slanted rays of the sun cause the chemical activity that is necessary to start the solvent flowing from the moth's tiny mouth. Or possibly the timing is done by temperature or by the amount of moisture in the morning air. In any case, you will have to watch your cocoons early in the morning if you do not want to miss the most wonderful part of the entire life cycle of the silk moth.

The moth pushes and climbs its way out of the cocoon. It looks wet, wrinkled, and bedraggled. When it is entirely free, it may rest for a while on its empty cocoon, or it may crawl or flutter to some other support, where it can wait for its wings to dry. It may just wait quietly, or it may move its small, crumpled wings back and forth. Gradually the wings will spread out. They will look a bit larger and stronger as blood goes from the insect's body into the veins of its wings. But the wings are still very small in proportion to the body. The entire wing span is only an inch and three quarters. The moths may flutter around a bit, but few, if any, will be able to fly. Their weak little wings cannot lift their heavy bodies.

Be sure to save the empty cocoons so that you can examine them later. You will discover why they cannot be used for silk.

Female silk moths laying eggs. *(Japan Silk Association, Inc.,* and *Japan Raw Silk Exporters Association)*

As you look at the newly emerged moths, you will notice at once that some are shorter and much fatter than others. The short, heavy ones are the females, each filled with hundreds and hundreds of eggs. The bodies of the males are thinner and slightly longer. You can also recognize the males by the way they behave. They approach the females with a queer kind of movement that is called courtship behavior, or "the love dance." Many kinds of animals have special dances they do in order to attract mates.

The mating occurs almost immediately after the moths have left their cocoons. Soon the mating is over. The males have then ended their part in the life cycle. In a few days, or perhaps even on the same day, the male moths die.

Each female, however, becomes busier than ever. Before evening of the day in which it has worked its way out of the cocoon and found a mate, it lays its first batch of eggs. One silk moth will lay from three hundred to eight hundred eggs. It will spend three or four hours laying eggs on the first day. It may lay more eggs during the next two or three days.

It is interesting to note that some of the very old books about silk moths say that the female lays two hundred to three hundred eggs. But from modern Japan comes word that today's silk moth often produces as many as eight hundred eggs. The years and years of careful breeding have improved the egg-laying records.

Watch the egg-laying process through a magnifying glass. The moth deposits each egg with its ovipositor, a special little organ located in its tail. If you place the moth on a clean piece of paper, it will lay its eggs neatly, one by one, on the paper. The eggs are pale yellow and sticky when they come out. They will stay right on the paper until they are scraped or washed off.

If you want to keep the eggs of each moth separate from other eggs, just put a large funnel over the female when it begins to lay. The moth will stay right there, quietly laying eggs, for one, two, or three days, until all of its eggs have been deposited. Then the moth may flutter around a few times. Soon it will die, for its part in the life cycle is completed with the laying of the eggs.

In earlier times in China and Japan, the eggs were laid on strips of paper or on pieces of cotton cloth. The silk farmer then sprinkled ashes over the new-laid eggs, rolled them up, and buried them in the ground to keep them cold during the winter. In the ground, the eggs were not cold enough to freeze but just cold enough to keep from hatching. When spring came and the mulberry trees began to show green leaf buds, the rolls of eggs were dug up. The ashes were shaken off, and the eggs were washed several times in cool water.

Today, in all countries where silk is produced, silkworm eggs are supplied by breeders, who use scientific methods to make sure that the eggs are from healthy moths. Since the dread silkworm diseases are passed on from the female moths to their eggs, and then on to the larvae, every egg-laying female is tested before the eggs are saved for a new crop of silkworms.

In the breeding plant, each female moth is put into a little

Silkworm eggs being weighed and packed for sale. *(Japan Silk Association, Inc.,* and *Japan Raw Silk Exporters Association)*

cotton bag to lay its eggs. The eggs stick to the cloth on the inside of the bag. When all of its eggs have been laid, the moth is taken out. The moth is then ground up and carefully examined through a microscope. The scientists who do this job are trained to recognize the germs of silkworm diseases. If a moth is found to have germs, its bag of eggs is destroyed. But if the moth has no disease germs, the bag that holds its eggs is turned inside out. The tiny eggs are scraped off the cloth and washed several times. Then they are placed on sheets of paper to dry. When they are dry, they are put into boxes, which are then stored in a cool, dry place.

When you buy silkworm eggs, you will probably get them in batches of a hundred. In this country silkworms are raised mainly as a hobby or for scientific study. A hundred eggs yield plenty of silkworms for these purposes. But in countries where silkworms are raised for commercial silk, the eggs are sold by the ounce.

Here are some facts about one ounce of silkworm eggs:

There are about 40,000 eggs to the ounce.

The silkworms from 40,000 eggs eat 3,500 pounds of mulberry leaves.

The cocoons from 40,000 silkworms weigh about 140 pounds.

These cocoons yield about 12 pounds of raw silk.

Thus, the silk farmer who begins with an ounce of eggs ends with twelve pounds of silk – but only after several months of day-and-night work by everyone in the family, including the children. It is amazing that silk cloth comes from a little insect. But what is even more amazing is that people for generations have devoted their skill, patience, and back-breaking labor to the raising of silkworms and the turning of the fat little cocoons into shimmering fabrics.

10: The Silkworm Reaches Europe

Until after 500 A.D., silk was made only in countries in Asia. All of the silk used in Greece, and later in the various parts of the Roman Empire, was brought in from Eastern lands. Silk continued to be rare, precious, and so expensive that some rulers forbade the wearing of pure silk garments, even by members of the noble families. Then something happened to change the history of silk. It happened in Byzantium some time in the sixth century A.D.

Byzantium is the ancient name for an important city in the land now called Turkey. Byzantium was later called Constantinople and, more recently, Istanbul. Byzantium-Constantinople-Istanbul is different from all other cities in the world. It is a gateway to two continents, Asia and Europe. It is located on the Strait of Bosporus between the Black Sea and the Sea of Marmara. For hundreds of years, products and ideas flowed through this gateway city from Eastern lands to the lands in the West. It is not surprising, then, that silkworms entered Europe by way of Byzantium.

At the time when it played a part in the history of silk, Byzantium was the capital of the Eastern Roman Empire. It was ruled over by the Emperor Justinian and the Empress Theodora. The silk merchants in Byzantium did a brisk business, bringing in silk from Persia (now called Iran), the land just east of Byzantium, and then selling it in Greece, to the west. They also shipped silk to Rome and to other centers in the Roman Empire. Much silk was used in Byzantium itself. In fact, the city had become famous in ancient times as a center for silk fabrics. In Byzantium it had long been the custom for weavers to unravel heavy Chinese silk fabrics, which were brought in from Persia. The unraveled threads were then rewoven on Byzantium looms, which turned out a sheer, light-weight silk cloth. This delicate silk was in great demand all over the Roman Empire.

When Justinian came into power, he felt that the silk merchants were making too much money, and so he made a law to control the price of silk. The merchants threw up their hands and raged against the emperor. They said that he was trying to force them to sell their silk for less money than they had paid for it. With the new law, they said, there was no more money to be made in silk. They threatened to go out of business.

Perhaps this was just what the Empress Theodora had hoped would happen. At any rate, she stepped forward and took possession of all of the silk then in Byzantium. The emperor himself took control of the silk business. Things went well for a while, until a long war with Persia cut off the supply of silk. So there they were, the emperor and the empress, in the silk business but without any silk and with no place from which they could get any.

The Emperor Justinian sent out word that he wished to find a new source of silk. Most of all, he wanted to find out how to produce some silk right there in Byzantium.

One day when Justinian was walking in his garden, a servant brought word that two ragged monks had appeared at the palace door begging to see the emperor. They said they had information that would gladden the emperor's heart. The servant did not know whether to admit the monks or to send them on their way. The emperor was weary from the news of the Persian wars and from troubles within Byzantium, but, with a sigh, he told the servant that he would see the two monks.

They were begging monks, the kind of dirty, footsore travelers who spent their whole lives wandering from place to place, begging for their food and sleeping by the side of the road. In their hands they carried canes to lean upon as they trod the rocky roads.

When they came into the presence of the emperor, they told their tale at once. They had heard, they explained, that it was the emperor's wish to produce silk. They had just returned from China, where they had lived for many years. While there, they had learned the secret of silk. And then the two threadbare and dusty pilgrims told the beautifully robed emperor about silkworms and mulberry trees and the lustrous cocoons that are unwound for silk.

The emperor was amazed and excited. Silk, the fabric for emperors, made by a tiny crawling creature! He must send at once for some of these little makers of silk! The monks then explained that it would be impossible to carry live silkworms across the long route from China. They also explained that the penalty for taking any out of China was death. But the emperor had caught a vision of his own land as a great producer of silk. He *must* have the silkworms regardless of risk, regardless of cost. He offered the monks a great reward if they would help him get started raising silkworms in Byzantium.

And so the two monks retraced their steps. They went back to China and there risked their lives by stealing enough silk-moth eggs to fill their canes, which they had hollowed out for the purpose. They made their return trip to Byzantium in the cold days of winter in order to keep the eggs from hatching.

As soon as they reached the court of Justinian, the monks set up a nursery for hatching and raising the silkworms. They found some black mulberry trees growing near the palace. The leaves from these trees were used until white mulberry trees could be grown, for, along with the eggs, the monks had carried seeds of the white mulberry.

Justinian put the two monks in charge of silk raising. He encouraged his subjects to plant mulberry trees and to raise silkworms. The monks were urged to teach as many people as possible how to produce silk. But the emperor kept the control of all silk in his own hands.

Beginning with very small amounts, silk was finally produced in Byzantium. For many years, though, it was more costly than the imported silk had been. But the important thing was that silk had made a start in Europe.

Slowly, very slowly, silk culture spread from Byzantium to Greece. Then the Mohammedans carried it to North Africa. From there the Moors brought silk to Spain and southern

Italy. After that, France, England, and the New World learned the secrets of silk. Silk was a mystery no longer.

Once silkworms and mulberry trees had escaped the secrecy of ancient Eastern lands, they became interwoven with the westward flow of civilization. They were carried from one land to another by conquerors in warfare, crusaders who journeyed eastward to the Holy Land, and adventurers who traveled as spies for their rulers. For hundreds of years, silk had a part in the exciting dramas of history.

There is a belief in France that the white mulberry tree became established there long before the French knew how to raise silkworms. In 1147 A.D., a knight named St. Aubon went on a crusade to Syria, one of the lands in the East. When he returned to France, he had with him a tiny white mulberry tree, which he had somehow managed to keep alive during his long trip. He planted the tree in his castle garden. It was a kind of souvenir of his travels. The remarkable thing is that the same tree was still growing more than six hundred years later! It was honored as the first tree of its kind to grow on French soil.

The crusaders brought many new things to France and to other countries of Europe. It was the custom during the Middle Ages for knights to leave their homes in Europe and journey great distances by land and sea to the Holy Land. They were not just making religious pilgrimages to a sacred place. No, they went there to fight — to drive out the non-Christians. It was often a tragic and bloody business. But those crusaders who escaped death in battle and survived the dangers of travel brought home exciting tales and wondrous treasures.

The crusaders from Europe found that the Eastern lands were inhabited by highly civilized people who had already developed art, science, mathematics, and other areas of learning far beyond anything then known in Europe. The crusaders brought home decorated leather and metal objects of beautiful workmanship; many kinds of spices, which greatly improved the flavor of European foods; and rich fabrics – velvets, brocades, and embroidered silks of such richness that the lords and ladies and their kings and queens were no longer satisfied with fabrics of linen and wool and cotton.

As early as 1200, there were mills in France where silk was woven. The raw silk, however, was not produced in France. As in the United States today, all of the raw silk had to be imported before silk cloth could be made. France used silk that was grown in Italy. This was a rather expensive arrangement, and the rulers of France tried several times to set up sericulture at home.

Finally, in 1603, Henry IV, king of France, did a number of things to get his people started producing silk. He urged farmers to plant mulberry trees and to rear silkworms. People tried to follow the wishes of their king. They planted mulberry trees. Silkworm eggs were brought in from Italy. The eggs hatched out. But then the troubles began. Many of the baby larvae got sick and died. Those that lived to spin cocoons lived to break out of their cocoons as well and so spoiled the silk. The silk growers were left not with the big profits they had expected, but with dead silkworms and pierced cocoons. The big trouble, of course, was that they did not know the facts about sericulture.

When King Henry realized that it would take more than

A grove of mulberry trees. *(Japan Silk Association, Inc.,* and *Japan Raw Silk Exporters Association)*

nurseries of mulberry trees and eggs of silkworms to give sericulture a start in France, he acted promptly. He sent to Italy for experienced silk growers.

There are stories, which may or may not be true, of some of the ways in which the French learned the secrets of producing silk. They sent representatives to Italy who disguised themselves and worked in the silk-rearing sheds. They made sketches and notes, which were smuggled back to France. They bribed Italian employees to leave their jobs and come to France to teach the French how to produce silk.

By one means or another, the French king was able to get some specialists to come to France. Little by little, the French began to raise their own silk. The entire silk project is said to have cost the French court great amounts of money. But, once the people had been given instructions, sericulture was successful in France. It flourished there for hundreds of years and paid back its original cost many times over.

French sericulture had one big setback, however. This was in 1685. At this time, Protestants were being persecuted in France. They were denied their religious freedom. Some 400,000 of them, the Huguenots, fled from France. It just happened that the Huguenots were the people who did most of the work in the silk industry. Many of them were weavers of silk in the famous French silk mills. When the Huguenots left France, there were only a few people remaining who could run the silk looms. It took France more than thirty years to recover its place as a leading producer of silk.

Meanwhile, the Huguenots spread across Europe, going to lands that allowed them to worship according to their own religious beliefs. They went to Germany, England, Switzerland, and other countries. With their knowledge of the silk industry and their skill in weaving, the Huguenots helped silk production in their new homelands.

According to early records, some silk was produced in England as early as 1251. Probably the weaving was done in England from raw silk that had been grown in Italy or Spain. England became famous for the weaving of special kinds of silk fabrics – tufted taffeta, figured velvet, and other fancy silks.

Up to the time of Queen Elizabeth I, every single pair of stockings had to be knitted by hand. Poor people, of course, never knew what it was to wear stockings. They just wrapped strips of cloth around their legs and feet or went barefooted. People who could afford yarn had hand-knitted stockings. In those times no lady would think of sitting by the fireplace in the evening without her knitting in her hands. Then, in 1589, a most important invention was made – a machine that would knit stockings.

Queen Elizabeth was greatly interested in the stocking machine. Personally, however, she was disappointed because the machine knitted only wool or cotton stockings. The queen preferred stockings of silk. Nine years after the machine had been introduced, its inventor was able to adjust it so that it could use silk thread. A pair of fine silk stockings was made and presented to the queen. They were, as far as we know, the first machine-made silk stockings in the world. Until nylon became plentiful, much of the silk spun by silkworms all over the world was knitted into ladies' stockings.

When we look back over the long history of silk, it is interesting to notice how many names of rulers are part of the story. First, Huang-Ti and Si-Ling-Chi, emperor and empress of China so long ago that their names and deeds are lost in legend. Then a Chinese princess married an Indian prince and carried sericulture to India. Emperor Justinian and Empress Theodora of Byzantium introduced silkworms to Europe. King Henry IV of France planted royal mulberry nurseries and sparked the silk business in France. And Queen Elizabeth I of England demanded, and received, a pair of silk stockings.

There were other rulers in other lands – kings and queens and princes and princesses – who encouraged the spread of sericulture among their people in order to secure for themselves some of the wealth that came from the production of silk. Sometimes they succeeded and sometimes they failed.

King James I of England was one of those who was not successful. He tried to give sericulture a start in the American colonies. And so, the story of silk moves to still another continent, and another chapter.

11: Silkworms in the New World

King James did not like tobacco, which was being shipped to England from the American colony of Virginia. He saw the people of his court smoking the brown dried leaves, and he made up his mind to put a stop to it. The best way he could think of was to cut off the supply. But this was not easy to do because the colonists in Virginia, with their huge tobacco fields, were dependent upon the English market. If England did not buy the tobacco, the colonists would lose their way of making a living.

England, since as early as the thirteenth century, had woven silk cloth. The silk mills were kept busy turning out beautiful fabrics made from raw silk that was produced in Italy and other southern countries. Many times, attempts had been made to grow silkworms in England. At one time, King James had a silk-raising project on land that is now covered by Buckingham Palace. This, like other projects, failed perhaps because of the cold and dampness.

When King James considered the problem of having to

import all of the raw silk used in England's silk mills and his great dislike for tobacco, he reached a logical conclusion. The Virginia planters should raise silkworms and mulberry trees instead of the hated tobacco! Accordingly, he gave some orders. The Virginia planters were told to stop raising tobacco. The king sent over mulberry seeds and eggs of the silk moth. He offered grants of money and land. He did everything he could to encourage the planters to produce silk, but the planters paid little attention to the offers or requests of their king, who was more than 3,000 miles away.

Next, King James made a law fining any Virginia planter about fifty dollars unless he grew at least ten mulberry trees for each hundred acres of his plantation. Again, the planters paid little or no attention. They kept right on raising tobacco, and men kept right on smoking it in the English court.

Thirty years later, the Virginia Assembly tried to encourage the production of silk. An award of ten thousand pounds of tobacco was offered to any planter who would send about a thousand dollars' worth of raw silk to England. Other awards were offered later, but none of them was ever claimed. The planters went right on growing tobacco. There was much money to be made from tobacco and very little, in comparison, to be made from raw silk.

England, of course, did not want the colonists to make silk cloth. She merely wanted them to produce the silk and then send it to England to be woven. The big profit would be made by the mill owners in England and not by the growers of silk in the colonies. All these things happened more than a hundred years before the colonies became free from England.

The colony of Georgia had a short history in sericulture. In the early seventeen hundreds, a clergyman was sent from England to teach the people of Georgia how to produce silk. Free land was given to settlers who would agree to plant a hundred mulberry trees on every ten acres. Free silkworm eggs and free mulberry seeds, as well as land, were offered to anyone who wanted to go into the business of producing silk. For a while, the silk projects in Georgia were successful. Three years after the arrival of the clergyman-teacher, a gift of eight pounds of raw silk was sent to the queen of England. There must have been quite a bit of raw silk produced in Georgia at that time, for soon a filature was built in Savannah. In a single year (1759) the Savannah filature reeled ten thousand pounds of raw silk from local cocoons.

But silk raising never became a really big operation in Georgia. Just as silk could not compete with tobacco in Virginia, so silk was overpowered by cotton in Georgia. There was less labor and more money in raising cotton than in silk.

Other colonies attempted sericulture. South Carolina, Connecticut, and Pennsylvania were among those to try. Various awards were offered to encourage farmers to raise silkworms and mulberry trees. In 1770, a filature for reeling cocoons was built in Philadelphia. It was not unusual for ladies to raise silkworms in order to get silk for their own use. They would take the cocoons to a filature and bring home the reeled silk. Then, on their own looms, they would weave cloth that would be made into elegant silk gowns.

Then came the Revolutionary War, which interrupted the production of silk in America. After the war, sericulture began again, on a very small scale, in various parts of the United States. And, then, suddenly in the 1830's there was a silk boom!

The boom was started by a new mulberry tree, *Morus multicaulis,* which was brought here from China by way of the Philippine Islands. This *Morus* was said to grow faster and to have more and larger leaves than the white mulberry. And it was said to be good for silkworms. To import the trees cost only four dollars a hundred. People began buying them and planting mulberry groves. New areas were mapped out to be the great silkworm farms of the future. As more and more people bought the imported mulberry trees, the price, naturally, went up. Within six years, *Morus multicaulis* trees were selling for as much as five hundred dollars a hundred! Men were getting rich on mulberry trees. They did not have to bother with the drudgery of raising silkworms.

Many people, of course, did go into sericulture. One early silk grower left a record of the way in which he began. First, he bought three hundred young mulberry trees. He planted the trees in the ground *horizontally,* six inches below the surface. He did this as an experiment. He had an idea that shoots would grow up from many places on each of the lying-down trees. And that is exactly what happened. In a very short time, 3,700 shoots were showing above the ground. Each shoot grew into a new mulberry tree. His experiment gave him 3,700 young trees instead of the three hundred he would have had if he had planted in the usual way. Within six weeks, he had enough mulberry leaves to feed 6,000 silkworms. And in ten weeks the shoots were two and a half feet high. (If you want to start a thick hedge of mulberry trees, it might be worth it to try planting a young tree on its side. It might work for you as well as it did for the silk grower in the early nineteenth century.)

For a while, *Morus multicaulis* caused a burst of activity in

American sericulture. Then something happened and the boom was over. First, there was a great financial crisis in the country, the panic of 1837. The raising of silkworms and mulberry trees, like all other kinds of business, was affected. At the same time, it was discovered that *Morus multicaulis,* the fast-growing, big-leafed mulberry tree that had been planted with such great hopes and enthusiasm, was not hardy enough for the American climate. It was a good tree in China and in the Philippines. But here, it grew for a while and then died. Those who had put their faith and money in the new mulberry tree lost out.

The final blow to United States sericulture happened in 1844. During that year a fatal tree disease hit all mulberry trees in the country. The trees were wiped out and with them, of course, all silkworms. The silkworm growers who had managed to survive other catastrophes now lost everything.

Sericulture never again got a real start in the United States. The main reason for this was finally recognized. It was not because of the climate. It is, silk experts believe, because of the *people.* Americans do some things by hand, it is true. But most things are done by machines. And most things are done fast.

There is no way to speed up the rearing of silkworms. There is no way to use machines to care for them. As you now know, rearing silkworms means hand care, hand cleaning, hand feeding. It is like taking care of hundreds – or thousands – of tiny babies all at once. The work of caring for the silkworms is done by families – fathers, mothers, and children of all ages. Much of the work is done in their homes. In some countries, during the feeding stage when the larvae are growing fast, the trays of silkworms have to be spread all over the house. People give up their bedrooms and even their beds to the precious silkworms.

When American silk growers discovered the back-breaking drudgery involved in producing a few pounds of raw silk, they lost heart and turned to other kinds of work.

There was, however, one interesting exception. It happened in California in our own century. In 1927, two brothers, Dale and Glen Hurst, joined forces with John Abdel Nour, an expert sericulturist from Armenia. Mr. Nour had been in the silk-producing business in several countries in the Orient. He and the Hurst brothers believed that silk could be produced profitably in California. They would use scientific methods and typical American mass production techniques.

First, they planted a grove of 400,000 mulberry trees. Around the edges of the mulberry grove, they planted Osage orange trees. They built a huge rearing shed for the silkworms, heat-controlled spinning rooms where the cocoons were to be spun, and a large filature to reel the silk. They imported their eggs from expert silk growers in France. The location they chose, San Marcos, has an ideal climate for silkworms and mulberry trees. There were always some trees in leaf. The hatching of the larvae was rotated so that there were always some insects in each stage. It is reported that two hundred and forty crops a year were raised at the San Marcos sericulture plant.

During World War II, the California plant turned out yards and yards of silk, which was used to make parachutes and silk bags for the explosive powder used in bombs. During the war years, there was, of course, no raw silk coming in from Japan and very little from any place else. The California-made silk was most important to the success of the war.

After the war, however, two things changed the outlook for silk production. Nylon had been invented, and parachutes, as well as other things that had once been made of silk, were now made of the new man-made fibers. And Japan went back to its age-old activity of producing raw silk. Not only did Japan take up silk producing where it had been interrupted by the war, but it began to turn out more silk than ever.

Today, if you write to the United States Department of Agriculture and ask them about raising silkworms, you will get a printed answer from the Agricultural Research Department. The heading on the paper is *Statement on Silkworms.* The first paragraph tells about the government research that was done on sericulture from 1884 to 1908. Since then, no further work has been done.

The second and final paragraph gives an important bit of advice. According to this bulletin from agricultural experts, anyone who is thinking of making a business of raising silkworms in the United States had better think twice. For many years, all attempts at sericulture here have met with failure. However, according to the last sentence in the government bulletin, anyone who raises silkworms as a hobby rather than a way of earning a living will gain most valuable experience.

12: Modern Silkworms in an Old English Castle

When Lady Hart Dyke of Lullingstone Castle was four years old, she became interested in silkworms. Later in her childhood she reared silkworms as a hobby. The more she learned about the fascinating little spinners, the more interested she became. In 1932 she decided to make silkworms more than a hobby. She set out to prove that good silk could be produced in England as well as in Japan and China.

Lady Hart Dyke read many books on sericulture. She went to Italy to study Italian methods of producing silk. She imported 4,000 mulberry trees from Palestine and several pounds of mulberry seeds from France. The mulberry trees were set out in orchards around Lullingstone Castle. The mulberry seeds were carefully planted in greenhouses until they grew into young trees. Then these were added to the acreage of mulberry trees.

Her husband, Sir Oliver Hart Dyke, designed special machines to unwind the cocoons. Some of the rooms of the castle were turned into silk-rearing rooms. One room became an incubation room with oil lamps to keep the eggs warm until the baby larvae hatched out. In several other rooms, the growing larvae were tended and fed. The old kitchen was used for a spinning room, where the mature larvae were placed to spin their cocoons. The castle laundry became an unwinding room, with kettles of hot water in which the cocoons could bob up and down as they were being stripped of silk.

When the silk farm was ready, Lady Hart Dyke imported thousands of eggs. The mulberry trees grew well in the green valley around the castle. There were plenty of leaves to feed the thousands of hungry silkworms. Within four years, the silk farm was a success.

Lullingstone — what a beautiful name for a castle! It is thus named for the soothing murmur of the Darent River as it slides over the stones. Lullingstone Castle is in the Darent Valley, in Kent, England. For five hundred years the castle has stood there in the quiet, peaceful valley. It is not surprising that silkworms, which are known to love quiet, did well at Lullingstone.

By 1937, the silk farm was turning out about five pounds of reeled silk each week. Most of it was from cocoons that had been spun there within the castle. The rest was from cocoons that had been raised by English school children, who sent or brought their cocoons to the castle to sell.

In 1937 there was great excitement at Lullingstone. The silk farm had been asked to produce twenty pounds of reeled silk to be woven into robes for the coronation of King George

VI! Twenty pounds – and only two weeks until the date of delivery! Lady Hart Dyke and her small staff of skilled workers had a busy two weeks. They worked in shifts around the clock. They unwound cocoons day and night. The skeins of shining silk were ready in time to be woven into robes of purple satin and velvet for the crowning of the king. How proud Lady Hart Dyke must have been to know that silk from English silkworms, reared on an English silk farm, reeled and woven by skilled English craftsmen, rested on the shoulders of the royal family!

During the early years of World War II, the quiet, busy life at Lullingstone Castle was not greatly changed. But in 1944 a most terrible kind of bombing began in the peaceful valley. Night after night the enemy planes flew low, shattering

Unwinding cocoons in the castle laundry. *(John Topham, Ltd.)*

the countryside with their blasting, roaring sounds and dropping death and destruction with their bombs. Gashes were torn in the fields, and homes were blown up. Each day brought its losses and each night its terrors.

Silkworms, as you know, are especially sensitive to loud noises. Now their short but busy lives were filled with noise. In the rearing rooms, the windows were continually being blown out. Each time they were replaced, another bomb would send them crashing to bits. Many of the silkworms – tens of thousands of them – died from the effects of blast. Others were cut to pieces by flying glass. The incubators were smashed. The machinery in the reeling room was put out of order.

Lady Hart Dyke fought to keep alive the silkworms that remained. Sometimes she slept in the rearing room with them. She would draw covers over the feeding larvae, keeping out some of the noise. Sometimes she herself had to duck for cover when the bombing drew especially close. But she protected her silkworms, determined that at least some of them would survive the horror that was falling upon the land.

One morning it seemed as though there could be no further hope. During the night, incendiary bombs (fire bombs) had been dropped around the castle. Every single mulberry tree had been burned black! Probably an ordinary person would have given up at this moment. But Lady Hart Dyke was no ordinary person. She needed food for her silkworms, and she needed it fast. She remembered that many English children had brought their home-grown cocoons to the castle. The children must know where to find mulberry leaves!

And so she sent out an appeal. She wrote an advertisement

Twisting the silk into skeins. *(John Topham, Ltd.)*

and had it run in fifty newspapers. This is what she wrote: "Help!! Help!! Lady Hart Dyke would be extremely grateful for sacks of mulberry leaves for her silkworms. Will those kind enough to help, please send their sacks by post or rail to Lullingstone Castle, Eynsford, Kent. Do please send as soon as possible after picking, as mulberry leaves ferment very quickly."

The children answered the appeal. Within two days, nearly a half ton of mulberry leaves had arrived at Lullingstone Castle. Children from all over England, and grownups, too, had sent boxes and bags of leaves, enough to save the silkworms.

After the war, the mulberry orchards were replanted and

the silk farm again flourished. Once more children carried or sent their boxes and baskets of home-grown cocoons to the castle to be unwound. Once more, English-made raw silk was produced at Lullingstone.

Just behind the castle is a very old church, St. Botolph's. It became the custom each Palm Sunday to hold a special ceremony in one of the small chapels – The Blessing of the Silkworms. For the ceremony, some of the silkworms and some skeins of raw silk were carried to the church. People who lived in the castle, the women who tended the silkworms and reeled the silk, and the people of nearby villages were very proud of the silk farm at Lullingstone. Its fame spread all over England. Thousands of visitors came each year. This sometimes made problems because many people make much noise. Visitors had to learn to be quiet and to go through the rearing rooms a few at a time.

In 1952, England was getting ready for the coronation of a new queen, Elizabeth II. At Lullingstone Castle, Lady Hart Dyke received word that Her Majesty wished the coronation robe to be woven from silk that had been produced at the Lullingstone silk farm. Once more the school children poured in with their cocoons, eager to contribute to the raw silk that would make robes for their queen.

The entire staff worked long and hard to reel enough of the finest silk. The reeled silk was then dyed purple and sent to the weavers. The royal velvet was woven by hand. Next, it went to the Royal School of Needlework, where expert needlewomen embroidered designs in gold on the fine silk velvet. This embroidered purple velvet was then used to make the Robe of State that Elizabeth II wore during the ceremony in which she was crowned Queen of England. Robe, velvet, silk – it all began inside the little silkworms at Lullingstone Castle.

Lady Hart Dyke had proved her point: good raw silk could be produced in England! Her silk farm had its own banner. On the banner were three green mulberry leaves, and on each leaf was a silkworm cocoon. Upon special occasions, the banner was carried proudly in processions.

Today there are no longer silkworms at Lullingstone Castle. Lady Hart Dyke has moved to another home in another part of England. But she is still interested in the production of silk. And here and there, where mulberry trees are plentiful, English children still raise silkworms as a hobby.

13:
Silkworm Gut from Spain

When a fisherman casts his dry fly or baited hook into a stream, it sinks below the surface. To the fish, watching it warily, the bait seems to be a tempting tidbit, floating freely in the water. The thin but strong cord that reaches from the bait to the fishline above the water is practically invisible. A fish will snap at the bait without knowing that it is fastened to a line that will yank it out of the water at the first good tug. The special cord with which fishermen tie their baited hooks to their fishlines is called a leader. For about a hundred years, the best leaders have been made from the glands of the silkworm, called silkworm gut. It is thin, tough, and nearly invisible in water.

For many years silkworm gut has also been used in surgery. Doctors have used it as thread to sew up wounds. The smooth, soft gut holds the cuts together until they are healed. The story of silkworm gut is another chapter in the history of the useful little silkworm. The setting for this chapter is Spain, in the province of Murcia.

Even before the beginning of the 1600's, Murcia was a well-known silk-producing area. Families raised silkworms and reeled silk, which was sold to merchants who came from all over Spain. Then, about a hundred years ago, a discovery was made that changed the silk business of Murcia.

One day some rag-picking gypsies were poking around in rubbish heaps in Murcia. They were looking for bits of cloth, paper, and other materials that they could sell. Suddenly one of the gypsies called to his friends. He had found a handful of dead silkworms in a broken bowl. They were larvae that had died just before reaching the spinning stage. Being worthless, they were quickly gathered up and thrown out by the silk farmer.

The gypsies, of course, knew about silk and silkworms. But they had never before seen any of them close, much less held them in their hands. They passed the dead insects around to be looked at. One of the gypsies tried in vain to squeeze out some silk. He was curious to see how and where the little insect could make silk threads, so, with his fingers, he pulled the silkworm apart. Inside, the fat, golden silk glands were clearly visible. He pulled out the silk glands and began to squeeze and stretch them. To his amazement, he found that they could be stretched to four – five – six times the length of the silkworm!

Soon the other gypsies were pulling the rest of the silkworms apart and taking out the silk glands. From them they made loops of strong, elastic silken cords. And this was the discovery that began the silkworm gut industry in Murcia.

Of course, like many stories of how something began years and years ago, this one may or may not be true. But it is the one that is told today in Spain. It is certain that *someone,* about a hundred years ago, discovered that a useful product could be made from the glands of dead silkworms.

Unfortunately, in order to use the glands, the silkworms have to be killed before they have a chance to spin. It is not possible to get both silk and gut from the same insects. And so, some of the families of Murcia stopped raising silkworms for the cocoons and began raising them for the gut. Perhaps there was more money to be made from the latter. Perhaps it meant less work. Perhaps both. In any case, Murcia has long been a world center for the production of fine silkworm gut.

For this product, the life cycle of the silkworm has to be interrupted at the end of the larva stage. When the larvae are just ready to spin, their silk glands are fully developed and filled with liquid silk. The silkworms are then killed and cured in an acid bath. The silk glands are removed and stretched to make commercial silkworm gut.

This work is done by girls and women, working in family groups. The women buy the eggs from breeders in the fall and keep them cool through the winter. In spring, the eggs are warmed between folds of cloth. Some women still carry the eggs in little bags, which they wear inside their clothes. Their body heat keeps the eggs warm till they hatch. A few years ago, however, the Spanish government helped some of the people of Murcia to get oil-burning incubators for their silkworm eggs. But many women still prefer the older ways.

On the first Friday in March, the eggs are carried to church to be blessed. Soon after this day, the baby larvae bite their way out of their shells. They are then fed on mulberry leaves, just as though they were being raised for silk. They eat and

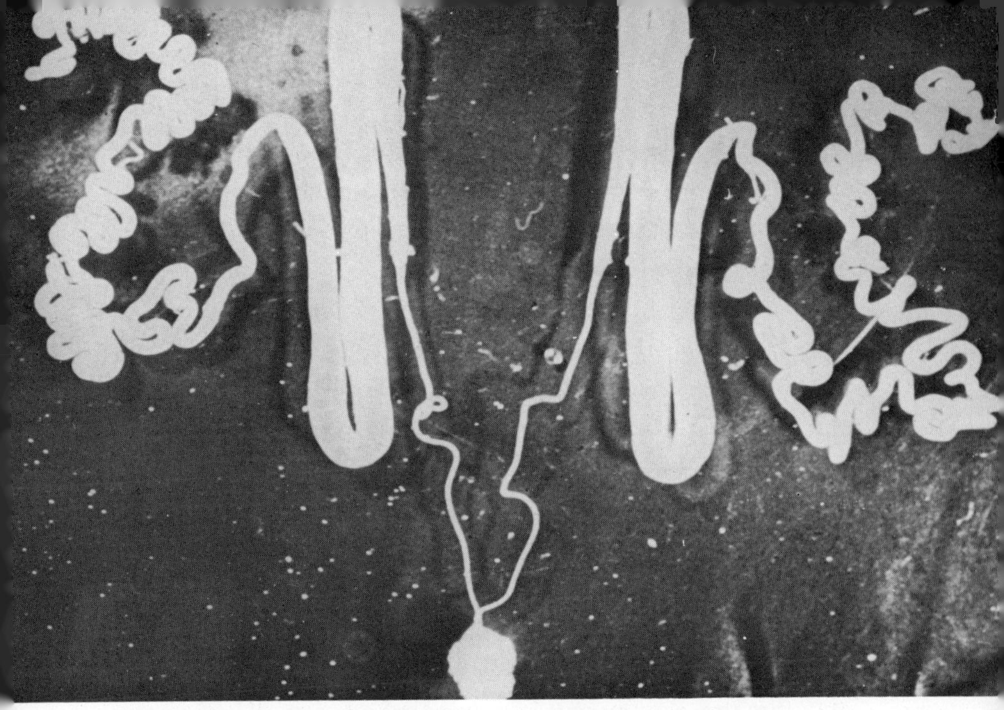

Silk glands removed from a silkworm. *(Japan Silk Association, Inc.,* and *Japan Raw Silk Exporters Association)*

grow and shed their skins four times. Then, when they are four or five weeks old, they stop eating and look around for supports on which to spin their cocoons.

At exactly this point, the silkworms are gathered up by the handful and dropped into the pickling bath, which is usually made of vinegar and salt. Some women, however, prefer vinegar and plain water, and a few use beer and water. The bath kills the silkworms and cures, or pickles, their bodies. The silkworms are left in the bath for at least twelve hours.

The opening of the bodies and pulling out of the silk glands is done by the elderly women. An old woman dips her hands into the pickling bath and picks out one silkworm at a time. She holds the larva with the fingers of her two hands and pulls it apart. Then her trained fingers pull out the two fat golden silk glands. She washes these and then stretches them as you might stretch a fat rubber band. From experience, she knows just exactly how far to stretch each pair of silk glands to make it as long as possible without breaking. One pair of glands may be stretched until it is as long as eighteen inches, and as thick as a heavy thread. The old women usually work outside in the sun with pails or basins of pickled silkworms at their feet. Clucking and pecking around them are the family chickens, who get the torn bodies of the silkworms as soon as the glands have been pulled out. Nothing is wasted in Spain.

When several hundred pairs of silk glands have been stretched, they are tied into bundles and washed. The washing is done by younger women, who kneel at the edge of a stream and dip the shimmering silk bundles up and down in the water. Between dips, the women comb the strands with their fingers. After the final washing, the gleaming bundles are hung on lines to dry in the sun. Finally, the bundles are opened and the strands of gut are counted and made into hanks. The glands from seven hundred silkworms make up one hank. When all the golden hanks are tied and counted, they are sent to a factory for the rest of the processing. The women of Murcia have finished their part for the year. In the fall, they will buy more eggs, and in the following spring, they will again go through all the steps necessary to grow and prepare silkworm gut for surgeons and fishermen.

The Spanish government has a sericulture station in Murcia. Trained workers, using four-foot models of silkworms, teach the women and girls from farms and villages how to raise silkworms and how to handle the valuable silk glands.

A special breed of silkworm is raised in Murcia. Each year some of the insects are allowed to complete their life cycle in order to provide enough eggs for the following year's crop. The cocoons they spin are larger and coarser than the cocoons spun by silkworms that are raised for silk. The Murcia cocoons are different in color, too. They are a delicate, pinkish yellow. These insects have long been bred for large, tough silk glands.

When the hanks of gut reach the factory, they are treated with boiling soapy water, which removes the yellow outer skin. They are then exposed to sulfur fumes, which bleach the gut until it is pure white. It is then polished and trimmed to make the smooth, tough, almost transparent material that you may have seen on the end of a fishline. The pieces of gut that are used by surgeons are coiled in bottles and covered with sterilized water.

In a good year, the women of Murcia have turned out as many as ninety million strands of gut, which were used by surgeons and fishermen all over the world. In recent times, however, there has been less demand for their product. Nylon has taken the place of silkworm gut. Most surgeons now use nylon thread for sewing up wounds. And many fishermen use nylon leaders on their fishlines. Nylon thread has the advantage of being any length and any exact thickness that may be needed. Gut, on the other hand, is only as long as the glands of a single silkworm and not uniformly thick. But on the end of a fishline in icy water, gut remains softer and sinks more easily below the surface. For these reasons, some fishermen still prefer gut to nylon. And so, the women of Murcia in Spain are still busy each spring hatching and raising their silkworms and sending hanks of golden strands to the factories.

If you are curious to examine the silk-making parts of a silkworm, why not take one apart as the gypsies did long ago in Murcia? As you raise your silkworms, you may find a dead one every once in a while. If a silkworm dies near the end of its growing period, put it on its side on a piece of cardboard. With a very sharp knife or a single-edge razor blade, cut the skin lengthways all around. Then peel back the skin. You will have no trouble finding the silk glands. In a mature silkworm, they are the biggest, fattest tubes in its body. Pull out the glands and stretch them slowly and gently. Wash them several times in cool water. Then dry them slowly in the air. Perhaps you can use your home-grown silkworm gut on your fishline.

If none of your silkworms die naturally, you may want to try the process used by the women of Murcia. Get a large glass or plastic jar. Fill it about three-fourths full of vinegar. Put in a fresh, unbroken raw egg. It will sink to the bottom. Then begin pouring in salt. Stir the vinegar to help dissolve the salt. Keep on adding salt until the egg floats. Then take out the egg. This is how some of the women in Murcia find out how much salt to use in their solutions.

Now, drop one or two silkworms into the pickling bath. Leave them there all night — for at least twelve hours. After that, you can pull open the silkworms and take out the glands, just as the old grandmothers do in Murcia. Then, if you have any chickens, find out if they find pickled silkworms a tasty tidbit!

14: Scientists Imitate the Silkworm

For thousands and thousands of years, man's clothing came from things that grew. Early man wrapped himself in winter in furs from the animals he killed. In summer he wore bark or leaves. American Indians clothed themselves in deerskin. Eskimos dressed in animal skins, which they chewed until they were as soft as cloth.

The art of weaving fibers into cloth began in prehistoric times. In the Stone Age, fibers from the flax plant were used to make fish nets. Later, flax was woven into linen cloth, one of the oldest woven fabrics. In ancient times, the soft fibers of the woolly coats of sheep were twisted, or spun, into yarn, which was woven into wool fabric. Cotton fibers were spun into cotton yarn for lighter, cooler cloth. The cliff dwellers in our own southwest discovered early how to shred the yucca plant and weave coarse cloth from its fibers. And in China, as you know, fibers from the cocoons of silkworms gave the world its softest, finest, and most beautiful fabric.

And so, man clothed himself for thousands of years – in

prehistoric days, in ancient times, and in modern times. He was always dependent on growing things. He needed fur, skins, and wool from animals. He needed plants like cotton, flax, and yucca. And, for beauty and luxury, he needed the little silkworm.

Sometimes the sheep would die, or there would come a year when the wool was not thick. Sometimes cotton crops failed. Sometimes, as you know, a crop of silkworms would be wiped out by a dread disease or because something had happened to the mulberry trees. When such things occurred, there would be a shortage of fabrics.

Not only was there always the problem of not being able to control the amount of fibers that could be produced, but also the fibers themselves were uneven in quality and size. Some were long, strong, and thick. Some were thin and weak. Some were both thick and thin. Fibers of uneven thickness were, of course, hard to weave. And so, after thousands of years of dependence on fibers that were grown by nature, men began to dream of *making* a fiber, a perfect fiber that could be completely controlled by man.

The first search was for something that would be as beautiful and fine and strong as silk, for silk, of course, was the most desired, most precious of all fabrics. This search went on for hundreds of years and, like the story of silk itself, took place in many different countries.

Before the middle of the 1600's, man could not look very deeply into the little things around him. He saw only what his eyes revealed to him. Then came the wonderful aid to seeing – the microscope. Through its lenses, men began to discover many things about plants and animals that had been mysteries. It is easy to imagine the great excitement with which a seventeenth century scientist would put small familiar things under his lens. Suddenly he would see, for the first time, the precise beauty of some scientific structure.

So it was with fibers of silk. Robert Hooke, an English scientist, studied raw silk through a microscope. He looked and looked, as you can do if you have a microscope handy. Even a small microscope today is probably more accurate and more powerful than the first microscopes back in the seventeenth century. Yet, through his early microscope, the English scientist got an understanding of the make-up of silk fiber. He wrote about it in 1664, saying he believed that man should be able to make a fiber like those made by the silkworm – but stronger and even more beautiful.

During the next 191 years, scientists in various countries talked and wrote about ways of making textile fibers. They tried out some of their ideas but were not successful in making fibers that could be woven into cloth.

Then, in 1855, a Swiss chemist named George Audemars was granted a patent for the manufacture of artificial silk. His chief raw material was cellulose, which is the solid part of all plant cells. Audemars used cellulose from the inner bark of the mulberry and other trees. He mixed the cellulose with other raw materials to make a thick, gummy liquid, which he then spun into fibers. He did this by dipping a needle into the liquid and drawing out a long strand or filament. The filament was then wound on a reel. He had the right idea but not the right method. His filaments were not strong enough to be used for cloth.

In 1883, Sir Joseph W. Swan, an English physicist, patented

a process that made it possible to manufacture long, strong filaments. But, unfortunately, the artificial silk made by the Swan process cost more than real silk.

Meanwhile, some important work was going on in France. In 1865, the French silkworms were being wiped out by the silkworm disease. And the great scientist, Louis Pasteur, was working on a cure. His assistant was a young man named Count Hilaire de Chardonnet. Count Chardonnet realized what a wonderful thing it would be for France if the great silk industry were not completely dependent on the delicate silkworm, so he began experimenting to find a substitute for real silk.

He started by making a careful study of the way in which a silkworm produces silk. And then he began working out ways to imitate the silkworm's methods. This may sound like a simple task, but it was far from that. It was a task that occupied Chardonnet for nearly twenty-five years!

Since he was copying the silkworm's methods, Chardonnet used mulberry leaves for his first raw materials. From these, he took the cellulose. Later, he got his cellulose by dissolving cotton in ether. And then, for many years, he experimented with different ways of changing the cellulose into liquid that could be forced through tiny holes in a spinneret. Warm air was blown on the soft, sticky threads, which became dry and shiny. Chardonnet was successful. His were the first man-made filaments that were completely practical for weaving into cloth. Twenty-five years after the French scientist had begun his search for a substitute for natural silk, his wife attended a ball dressed in a gown made from the shiny, new, artificial silk.

It was not until 1911, however, that artificial silk was manufactured in the United States. This was done by a small British firm, which has since grown into a huge company, The American Viscose Corporation. And not until 1924 was the artificial silk given a name of its own – rayon. Since its invention, rayon has been greatly changed and improved.

Two main kinds of rayon fibers are made: acetate and viscose. They are both made from cellulose, but different chemicals are used to change the cellulose to the syrup-like liquid from which the fibers are spun. When viscose rayon fibers come out through the spinneret, they pass through an acid bath, which hardens them. The fibers of acetate rayon harden in air.

Liquid being forced through tiny holes in a spinneret – the birth of man-made fibers. *(E. I. du Pont de Nemours & Co.)*

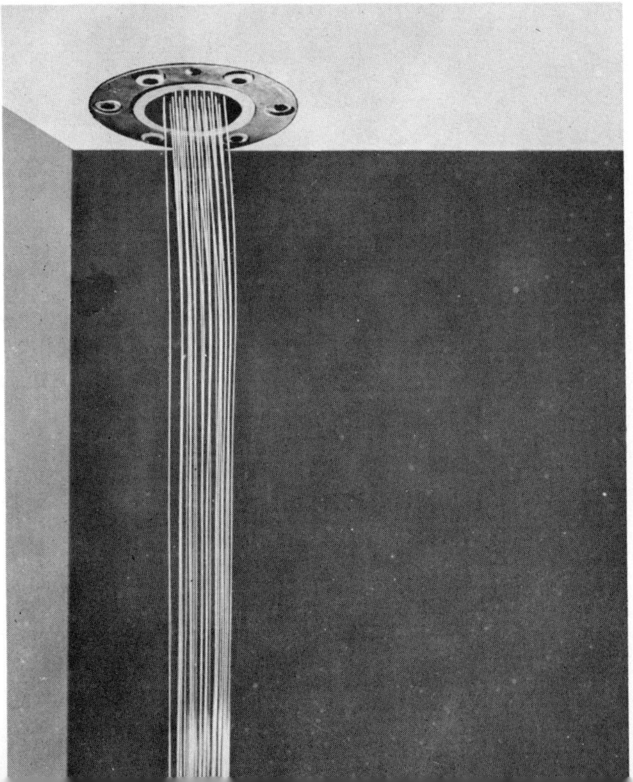

In 1951, the United States Federal Trade Commission ruled that acetate rayon was to be known simply as acetate. The commission also ruled that viscose rayon was to be called rayon. You may notice "acetate" and "rayon" if you read the labels on some of your clothing.

Today, acetate and rayon are much more than substitutes for silk. They are used to make fine clothing, rugs, upholstery, draperies, and even tough things like the cords in tires.

Rayon and acetate fibers are made by man. But they are made from natural materials, materials that grow in plants. And so, perhaps we should not call them "man-made," for the raw materials are provided by nature just as mulberry trees give the silkworms what they need in order to make silk. Scientists in England, Switzerland, France and the United States, working over a period of about three hundred years, finally learned how to imitate the silkworm.

15:
Test-Tube Fibers

In contrast to rayon, which is not a completely artificial material, nylon is both man-made and synthetic. That is, it is made by building up chemical elements and compounds into a new product. The raw materials in nylon do not grow in nature. They are chemicals that are taken from water, air, and coal or petroleum. The story of the discovery and development of nylon should interest anyone who is fond of chemistry. Nylon was discovered by a chemist who was not looking for it! It happened this way.

In 1927, the Du Pont Company began a program of special research. They were not searching for a new fiber, nor for any new products. They were doing something called fundamental research. This means research that is aimed at discovering some of the basic laws of science.

One phase of the fundamental research was headed by Dr. Wallace Carothers, a chemist. He studied molecules, the tiny, tiny particles that make up all matter. He was interested in the way in which some molecules join together in groups to

Here, greatly simplified, are the steps in making nylon:

1. Chemicals (from water, air, and coal or petroleum) are combined to make nylon salt.
2. The nylon salt is dissolved in water.
3. Some of the water is evaporated out in a huge dryer.
4. The salt is then put into a big pressure tank. Heat drives off the rest of the water. Heat and pressure then force the tiny molecules in the nylon salt to join together and make linear superpolymers. The material is now hot and molten.
5. The hot liquid is forced out through very tiny holes in a spinneret.
6. As the material comes out of the tiny holes, it hardens instantly in the air. Air changes it from a hot liquid to a cold solid. From each of the holes comes a very fine thread, or filament.
7. Several filaments are combined to make a strand of nylon.
8. The strands of nylon are stretched and drawn out between rollers. The stretching and rolling cause the molecules to line up in a special way that makes the strands strong and elastic.
9. The nylon thread is reeled on spools or bobbins.
10. The nylon thread is then woven into smooth, silklike fabrics.

Nylon is not the only test-tube fiber made today. There are Dacron, Orlon, Dynel, Acrilan, Saran, Vicara, Ardil, and others. Some are made by the Du Pont chemists and some by other chemical companies. Each new fiber has certain special

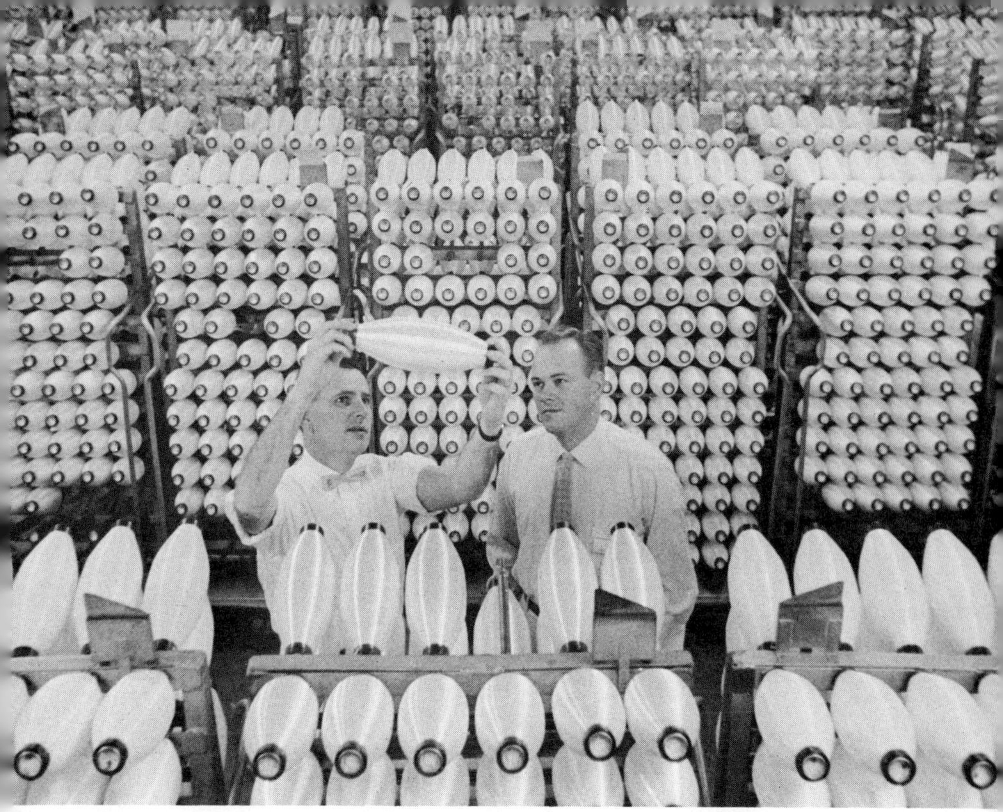

Nylon to be made into ladies' stockings. *(E. I. du Pont de Nemours & Co.)*

characteristics, but scientists are still experimenting. They are still trying to discover the perfect fiber.

Synthetic materials have greatly added to man's comfort and convenience. They have qualities that no natural fibers possess. They can be manufactured in great quantities. The amounts do not depend upon nature but are entirely within man's control. Fabrics made from synthetic fibers have become inexpensive and plentiful in most places in the world. Now some people make the mistake of thinking that these fabrics have taken the place of silk in the life of man, but this is not so.

Synthetic fabrics are better than silk for certain purposes.

They are called "miracle fabrics," and so they are in many ways. But they are not — and they are not intended to be — substitutes for silk. Nylon has its own special values. But silk is still silk. And for beauty and luxury and for some other qualities you will find out about when you do some scientific testing, nothing takes the place of real silk as made by the busy little silkworm.

16:
Collecting and Testing Silk

Many things that are made of fabrics have labels that tell the kinds of fibers used. Have a label hunt around your house. You can find out what fibers are in the fabrics you wear and use. Here are some places to look for labels: neckties, shirts, dresses, coats, suits, cushions, table cloths, upholstery, and rugs. You will find things made of nylon, Dacron, Orlon, rayon, cotton, linen, wool, and silk. Some labels will show that the material has been made of several kinds of fibers – silk and wool, silk and Dacron, cotton and nylon, and other combinations. And some labels will show that the material is pure silk, all wool, all linen, or 100 per cent nylon. Ladies' stockings today are usually all nylon. Socks, on the other hand, are often made of mixed fibers. New socks have the fiber content printed on the bottom of the foot. It is interesting to find out what kinds of fibers make up the things you wear.

You may wonder why, with all the wonderful man-made fibers, silk is still used. Silk has been studied for centuries, and its qualities have been compared with those of other natural fibers: wool, cotton, and flax (linen). More recently, scientists have been comparing silk with nylon, rayon, and other man-made fibers. Below are the characteristics of silk fibers as they have been summarized by some of the scientists who test fibers.

STRENGTH AND LIGHTNESS. Silk is the strongest of all natural fibers for its weight. A strand of raw silk is nearly one third as strong as a strand of the best iron wire of the same thickness. It is the lightness and strength of silk that made it the ideal material for parachutes until nylon was developed. Nylon is stronger than silk. Rayon, however, is less strong. Rayon is especially weak when wet.

LENGTH. Silk is the longest of all natural fibers. Man-made fibers, of course, can be made as long as desired. Threads made of long fibers are usually stronger than those made of short fibers.

ELASTICITY. Pure silk is very elastic. When it is pulled out of shape, it springs back again. When silk fabric becomes wrinkled, it goes back easily to its original smoothness. Next to wool, silk is the most elastic of the natural fibers. Nylon, however, is more elastic than any of the natural fibers.

HEAT CONDUCTIVITY. Silk is a nonconductor of heat. In winter, silk garments are warm because they hold in the body's heat. And in summer, silk feels cool because it keeps out the heat from the outside. Silk, therefore, is said to be a good insulating material. Nylon, too, is a good insulator.

WATER ABSORPTION. Silk soaks up, or absorbs, moisture. In normal atmosphere, the absorption rate of silk is 11 to 12 per cent. Nylon, on the other hand, absorbs only 4 per cent. Silk absorbs perspiration from the skin and then allows the moisture to evaporate from the fibers. This is a cooling process in the summer when people perspire most. Nylon, being far less absorbent, does not cool the body in hot weather.

AFFINITY TO DYES. Silk, unlike any other fabric, has an affinity, or kinship, to most kinds of dyes. Through the centuries, silk has been dipped, painted, printed, stamped, boiled, and treated to many other color processes. Vivid, lustrous colors and beautiful patterns have distinguished silk through the ages.

LUSTER. Silk is the shiniest of all natural fibers. Rayon, nylon, and other man-made fibers can also be made with a silklike shine. But pure silk has a rich, deep, glowing sheen that comes from the chemicals in the glands of the silkworm. This beautiful luster has not yet been duplicated in test-tube fibers.

TEXTURE. The shine of silk is due in part to the great smoothness of its individual fibers. Silk is smooth and soft to touch. Because most silk has no roughness, it does not soil easily. And when it does become soiled, the dirt is easily removed by washing or dry cleaning. No other natural fibers have the smoothness of silk. Nylon, however, is also smooth and washes even more easily than silk.

As you read the above characteristics of silk, you will realize that today there are man-made fibers that equal or surpass silk in many ways. And so, many fine things that were once made only of silk are now made of test-tube fibers, especially nylon. Yet nothing has replaced silk as a fabric of lustrous beauty and smooth, shimmering texture. For beauty, silk is still the queen of fabrics.

Then the two fibers came out through a single hole and stuck together to form one filament, or strand. You will notice the special shine of the fibers, the shine that gives silk its characteristic luster. Unravel some bits of cotton, wool, linen, and nylon. Compare the different fibers. You can learn to recognize silk.

FEELING. Silk cloth feels smooth and soft.

HEARING. Silk makes a little crackling sound when you rub two pieces together or when you squeeze a large piece of silk together in your hand.

BURNING. You will need to have your parents' permission to do the burning tests. Make certain that they are nearby and that you have a fire extinguisher handy. Do these tests in a safe place, away from curtains and other things that might catch fire.

Light a candle that is securely standing in a safe holder. Put the candle in the middle of a metal pie pan. Have ready some pieces of fabrics to test. Use small pieces, not more than an inch square, and use pieces of different kinds of cloth: cotton, linen, wool, nylon, rayon, and several kinds of silk. You will also need a pair of tongs or an ice pick.

To test a piece of fabric, hold it in the tongs or stick it on the end of the ice pick. Hold the edge of the fabric in the candle flame. As soon as the fabric catches fire, lay it in the metal pan and watch what happens. Different fabrics burn in different ways. One way of identifying them is by their burning characteristics.

Cotton burns fast with an orange-yellow flame. As it burns, it smells like burnt paper. It leaves a crisp black or gray ash.

A piece of cotton does not usually go out till it is all burned up. In other words, it is not self-extinguishing. This is one of its characteristics.

Linen burns in much the same way as cotton, but it gives off a smell like burning grass. It makes sparks and crackles slightly as it burns. When it is removed from the candle flame, it smolders and goes out. It is self-extinguishing. It leaves a delicate gray ash.

Wool burns with a yellow flame and the odor of burnt hair. When removed from the flame, it gives off clouds of blue-gray smoke. It is self-extinguishing. On the edges you will find swollen black beads, like tiny bits of coke.

Viscose rayon burns like cotton and gives off the same smell of burning paper.

Acetate rayon flames up and then melts. It leaves a hard black ball when it cools.

Nylon does not really burn. It melts, making bubbles. It turns to a thick light brown or greenish liquid, which hardens when it gets cold. It is self-extinguishing.

Orlon catches fire easily and burns with a yellow flame and a sweet odor. The smoke is black. It is not self-extinguishing.

Dacron does not catch fire easily. It melts when it becomes very hot. When it does burn, it has a yellow flame and a sweet odor. It is self-extinguishing.

Pure silk catches fire easily and burns steadily with an orange-yellow flame and an odor of burnt hair. It sizzles and sparkles as it burns. It is self-extinguishing.

Weighted silk (like taffeta, for instance) burns like pure silk but leaves behind a network of black ash from the metallic chemicals with which it was weighted.

You can do another kind of burning test by putting tiny pieces of fabric on the metal pan and putting the pan over a burner on the kitchen stove. Some of the fabrics will scorch and then turn black but keep their original shape. Others, the test-tube fabrics, are more likely to melt and then bubble and boil. You can melt a piece of nylon stocking in this way and see the thick greenish-brown liquid from which the nylon fibers were spun.

If you are interested in chemistry, you will want to find out more about different kinds of fibers and the work of chemists who make them, improve them, and test them. Today there are many science careers for men and women in the textile industries of the world.

17:
Sericulture in Modern Japan

For thousands of years, silk was a mysterious fabric – beautiful, greatly desired, and almost priceless. It is still beautiful and greatly desired. But today it is less mysterious and far, far less expensive than ever before. Silk costs less to buy, and there is more of it. Yet every inch of silk fiber is still spun by a mulberry-fed silkworm that has been reared and cared for like a baby. Where on earth does all of today's silk come from? And how do we get so much of it that practically everybody can wear something made of silk?

For the answers, we must visit the little farms that lie scattered all over the mountainous islands that form Japan, for it is on these little farms that most of today's silk is grown. Nearly all of the silk you wear and see in stores came from silkworms that were raised in the farm villages of Japan.

Japan is a small nation with a large population. The country is made up of four main islands and more than three thousand small islands. Much of the land is mountainous. To support its large population, Japan has always had to use its land wisely.

Silk farm in Japan. *(Japan Silk Association, Inc.,* and *Japan Raw Silk Exporters Association)*

The low areas that can be flooded easily are used as rice fields. Wheat and barley are planted on the higher level places. And the terraced mountainsides and steep-sloped hills are planted with fruit trees and mulberry trees. The crops of mulberry leaves support Japan's huge silk industry.

More than 760,000 Japanese farm families raise silkworms. For many of these families, sericulture is a part-time activity. They raise rice and other crops as well. But for some families, the production of cocoons is the main work and the main source of income. Much of the work done on these family silk farms follows the age-old pattern that the Japanese learned in the fourth or fifth century from the four little girls from China.

Except for its most northern islands, Japan has an ideal climate for sericulture. It has a long growing season. Most silk farmers have learned how to raise three crops of silkworms a year. One crop hatches out in May, when the mulberry trees begin getting their first leaf buds. One hatches in summer, and the third in the fall. The family on a small farm can make more money from their three crops of cocoons a year than from a single crop of wheat or rice.

And so, scattered across Japan's island nation are the families on little farms that produce most of the silk in the world today. More than half of the silk exported by Japan comes to the United States. The silk in your father's ties, in your mother's dresses, in scarves, suits, draperies, handbags, and other fine things probably came from silkworms raised by farm families in Japan.

Raising thousands of silkworms at a time takes much work. But it is not all heavy work. Some of it can be done by women and children and by old people who are no longer able to do other kinds of farm work. Every person in the family, except the newest baby, is a worker. The trees must be tended, the leaves gathered and chopped up. The larvae must be fed, moved, cleaned. The spinning time must be prepared for. And after the cocoons have been spun, they must be carefully sorted and packed in baskets, ready to be sold.

Today, most Japanese cocoons are taken to silk-reeling factories, or filatures, for unwinding. In some farm homes, however, there are little filatures where cocoons are still unwound by hand. Plenty of patient, skilled workers, such as the farm families supply, is perhaps the most important single reason for Japan's continuous success in the ancient art of sericulture.

So important is the production of silk that Japanese farmers spend much time and money improving their mulberry trees. Special fertilizers are added to the soil to make the trees grow well. Good trees are those that have many good leaves. And good leaves are those that give the silkworms large quantities of the chemicals needed for spinning good silk. So, in Japan, good silk begins with good soil.

An acre of mulberry trees provides enough leaves to feed at least 160,000 silkworms a year. Many farmers divide their fields of trees into three sections, one to feed the spring crop of silkworms, one for the summer crop, and one for the fall crop. When the trees are allowed to grow naturally, they become tall. Since this makes hard work for the girls and women who gather the leaves, many farmers cut back the branches each year, clear back to the roots. Then new branches, like slender switches, grow up from the old roots each spring. These switches bear thick growths of leaves, which never are beyond arm's reach.

In May, life on a silk farm becomes very busy. The men and women wash the mulberry baskets and silkworm trays. They scrub out the rearing rooms. They spray everything with chemicals that will protect the silkworms from disease. When everything is sanitary and in perfect order, the farm family is ready to buy a little box of silkworm eggs. The eggs are bought from a licensed breeder, who is permitted by law to sell only the eggs from healthy moths. The farmer knows that the eggs he buys will hatch and grow into larvae that will spin good silk. Laws are strict about silkworm eggs because the silk industry is very important to the prosperity of Japan.

Getting ready for a new crop of silkworms. *(Japan Silk Association, Inc.,* and *Japan Raw Silk Exporters Association)*

The silk farmer brings his eggs home and spreads them on trays to hatch out. The room in which the eggs are kept is warmed, day and night, by little oil stoves or by charcoal burners. When the larvae have hatched out, the women and children begin to feed the baby silkworms.

Today, more and more silk farmers are starting their crops, not with eggs, but with new-hatched larvae. Scattered across Japan are numerous large incubation stations where silkworm eggs are hatched under ideal conditions.

Hatching begins early in the morning, usually just at daybreak. Light seems to be a very important factor in the hatching process. In the incubation rooms, artificial lights are turned on because, with the extra light, the hatching is speeded up. By ten o'clock all of the larvae are out of their shells, and by eleven o'clock the first feeding is given.

The new-hatched larvae are then rolled up with some finely chopped mulberry leaves, wrapped in paper, and sold to the silk farmers for rearing. The farmers carry the rolls of tiny larvae to their farms, spread them out on woven straw and bamboo trays (called "rearing seats"), and begin the feeding routine.

As you know, silkworms eat continuously. In order to feed and care for the growing larvae, every member of the family works hard from daylight to dark. As soon as the dew of early morning has evaporated, whole families are out picking the fresh green leaves.

The leaves are carried from the mulberry fields in deep baskets, which the girls bear on their backs. In the rearing shed, the leaves are chopped or ground fine before they are fed to the baby silkworms. The leaves are put through something that looks like a big food chopper. Some of the leaves are carefully stored for use during rainy weather. Silkworms will not eat leaves that are wet, wilted, or dead. The leaves cannot be stored for very long at a time because they ferment easily. If the leaves are too damp, they must be carefully dried off. If they are dusty, they must be both washed and dried.

At first the larvae are fed five times a day and twice at night. Members of the family take turns getting up or staying up to take care of the night feedings. As they grow bigger, the larvae eat larger and larger pieces of mulberry leaves. After they have shed two or three times, they can eat whole leaves. In their final growth stage, silkworms can eat very large leaves

Picking mulberry leaves. *(Japan Silk Association, Inc.,* and *Japan Raw Silk Exporters Association)*

and even parts of stems. As the silkworms eat more and more, the women and girls who gather the leaves have more and more work to do, cutting the mulberry leaves and carrying the heavy baskets back to the rearing sheds.

As the silkworms eat, they grow rapidly. Each day they are moved to give them more room. A net is spread over the tray. Pieces of mulberry leaves are scattered thickly over the net. The hungry silkworms climb up through the open spaces of the net to reach the fresh green leaves. As soon as enough silkworms have climbed up, the net is lifted off the tray and placed in a clean tray. Another net and more mulberry leaves

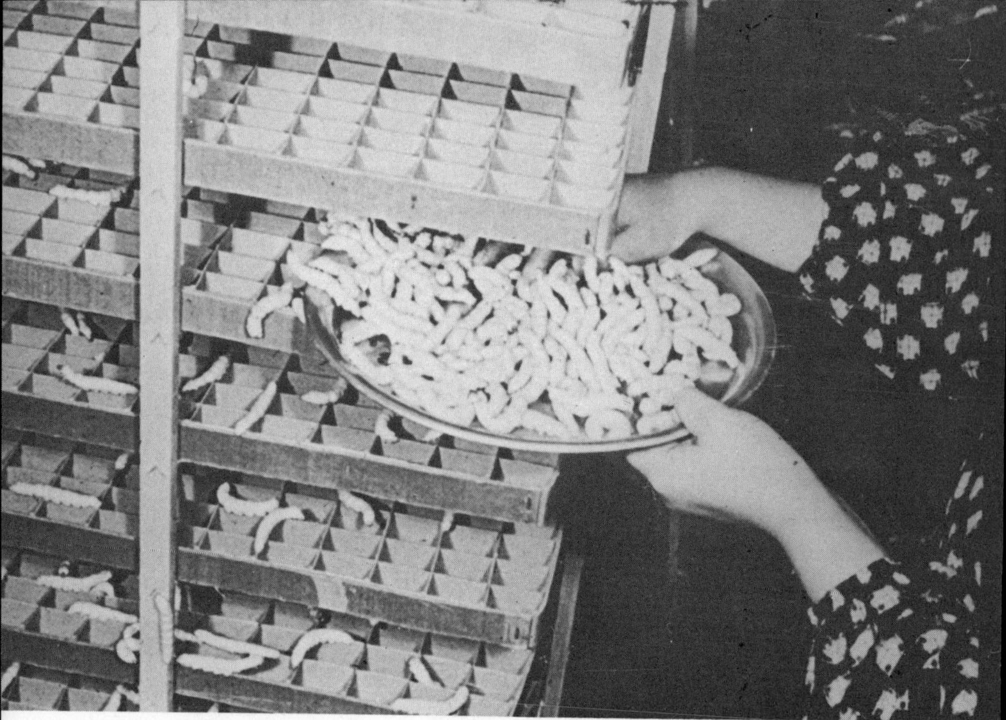

Putting silkworms into spinning mounts. *(Japan Silk Association, Inc.,* and *Japan Raw Silk Exporters Association)*

are then used to transfer the rest of the silkworms to a fresh tray. In this way, the larvae are continuously being spread out as they grow and, at the same time, are given clean trays daily.

The trays fit into stacks. For feeding, each tray is pulled out like a drawer and fresh leaves are scattered over the hungry silkworms. Cleaning the trays, in rearing sheds where thousands of silkworms are reared at one time, is a never ending job. Girls and women in clean white coats keep lifting, cleaning, and replacing trays all day long. Even then, the smell from thousands of silkworms is anything but pleasant.

As the silkworms eat, biting and tearing into the leaves, they make a sound like a downpour of rain on a tin roof. Some silk farmers who have unusually large rearing sheds say that their silkworms sound like the rushing of a mountain stream as they bite, chew, cut, and tear – day and night except for the periods of molting.

When the silkworms have become fully mature and are ready to spin, they are picked up by hand and gently placed, a handful at a time, on a cardboard spinning nest, where each silk-heavy larva finds its way to an empty little cubicle. There it soon spins the supporting lines of coarse silk from which its cocoon will be hung. Then, for about sixty hours, it spins continuously until all of the silk in its glands has been spun out. During this time, the silk farmer tries to keep the room as quiet and dry as possible. Each of the cardboard spinning nests, or mounts, has spaces for 196 cocoons. The mounts are

Mounts in a spinning room. *(Japan Silk Association, Inc.,* and *Japan Raw Silk Exporters Association)*

stacked or hung in rows to allow the free circulation of warm, dry air. If the cocoons become damp, there is danger of the silk being ruined.

When the cocoons are finished, the women, children, and old people of the family pull them out of the cardboard mounts and rub off the coarse, silky fuzz that is on the outside of the cocoons. The silk farmer then sorts his cocoons according to size, shape, and quality, and takes them to a silk cooperative. This is an organization that receives cocoons from many farmers and sells them to a filature. The price each farmer gets for his crop of cocoons depends on the quality as judged by officials at the filature.

In the filature, the cocoons are placed at once in huge drying machines. Hot air kills the pupae and takes out all moisture. This reduces the cocoons to 40 per cent of their original weight.

Drying and packing cocoons in a filature. *(Japan Silk Association, Inc.,* and *Japan Raw Silk Exporters Association)*

Grading cocoons in a filature. *(Japan Silk Association, Inc.,* and *Japan Raw Silk Exporters Association)*

The dried cocoons are packed in bags and placed in an air-tight storage room. They may be kept there for days, months, or even years before they are taken to the reeling rooms for unwinding.

The reeling process begins with the boiling of the cocoons. They are placed in a special machine, which holds the cocoons in a boiling bath for ten to twelve minutes. This dissolves some of the sericin, and the cocoons become soft. Fine filaments of silk, loosened from the cocoons, float on top of the boiling water. To unwind a cocoon, the end of its filament is lifted up and put onto a reel.

A single filament is too fine to be used as a textile fiber. For this reason, filaments from several cocoons, usually five to eight, are unwound together to make a single silk fiber. Since the various cocoons have filaments of different lengths, no two cocoons end at the same time. So, when the end of one filament is reached, a girl with sharp eyes and nimble fingers quickly reaches into the hot water and lifts up a filament from a new cocoon. In this way, the silk fiber is kept uniform in size.

In the most modern filatures, automatic reeling machines are now being used. The cocoons are still placed in boiling water to loosen the filaments. But the loose ends are picked up by the machine's finger-like parts, which lift them, a filament at a time, to the reels as needed.

After the first reeling, the raw silk is wound onto larger reels. While this re-reeling is going on, trained workers watch the fibers, checking them for weak or discolored places. The raw silk is taken off the large reels and twisted into shiny skeins. Thirty skeins are bound together into a bundle, which is called a book. Twenty-nine books are packed into a bale, which weighs 132 pounds. Skeins, books, and bales are units of measure in the raw-silk business. From the filature, bales of silk go to weaving mills in Japan and in other countries all over the world, especially the United States.

When you remember that the silk farmer has to raise 40,000 silkworms in order to produce twelve pounds of silk, you realize that each bale of raw silk is the work of 440,000 silkworms!

No wonder, then, that each farmer hastens to begin raising a second crop of silkworms as soon as he has been paid for his first harvest of cocoons. Once more, baskets, trays, boxes, and rooms are washed, scrubbed, and sprayed to make them sanitary. Then the farmer buys a new setting of government-inspected eggs to begin his next crop of silkworms, or he buys a roll of new-hatched larvae from an incubation station. And again everybody in the family works together gathering mulberry leaves and caring for the growing silkworms. And after the second crop, most silk farmers produce a third.

And so, from early May until late in the fall, the fathers, mothers, grandparents, and children on the silk farms in Japan are busy day and night. It takes a great deal of work by many people of all ages to raise the silkworms that produce the millions of little cocoons that supply the whole world with silk.

Where to Buy Silkworm Eggs

H. A. Dahl Co.
Biological Supplies
2017 Second Street
Berkeley 10, California

General Biological Supply House
8200 South Hoyne Ave.
Chicago 20, Illinois

New York Scientific Supply Co.
28 West 30th Street
New York 1, New York

Quivira Specialties Co.
Biological Supplies
4204 West 21st Street
Topeka, Kansas

Note: There are many other scientific supply companies in different parts of the United States. Many of them would like to sell silkworm eggs if they knew where to buy the eggs wholesale. If you raise silkworms as a hobby, you may be able to build up a little business for yourself at the same time – supplying fertile eggs to companies that deal in biological supplies.

Index

Numbers in italics indicate illustrations.

Cooper, Elizabeth
K.